I0797505

Floraphile's Almanac

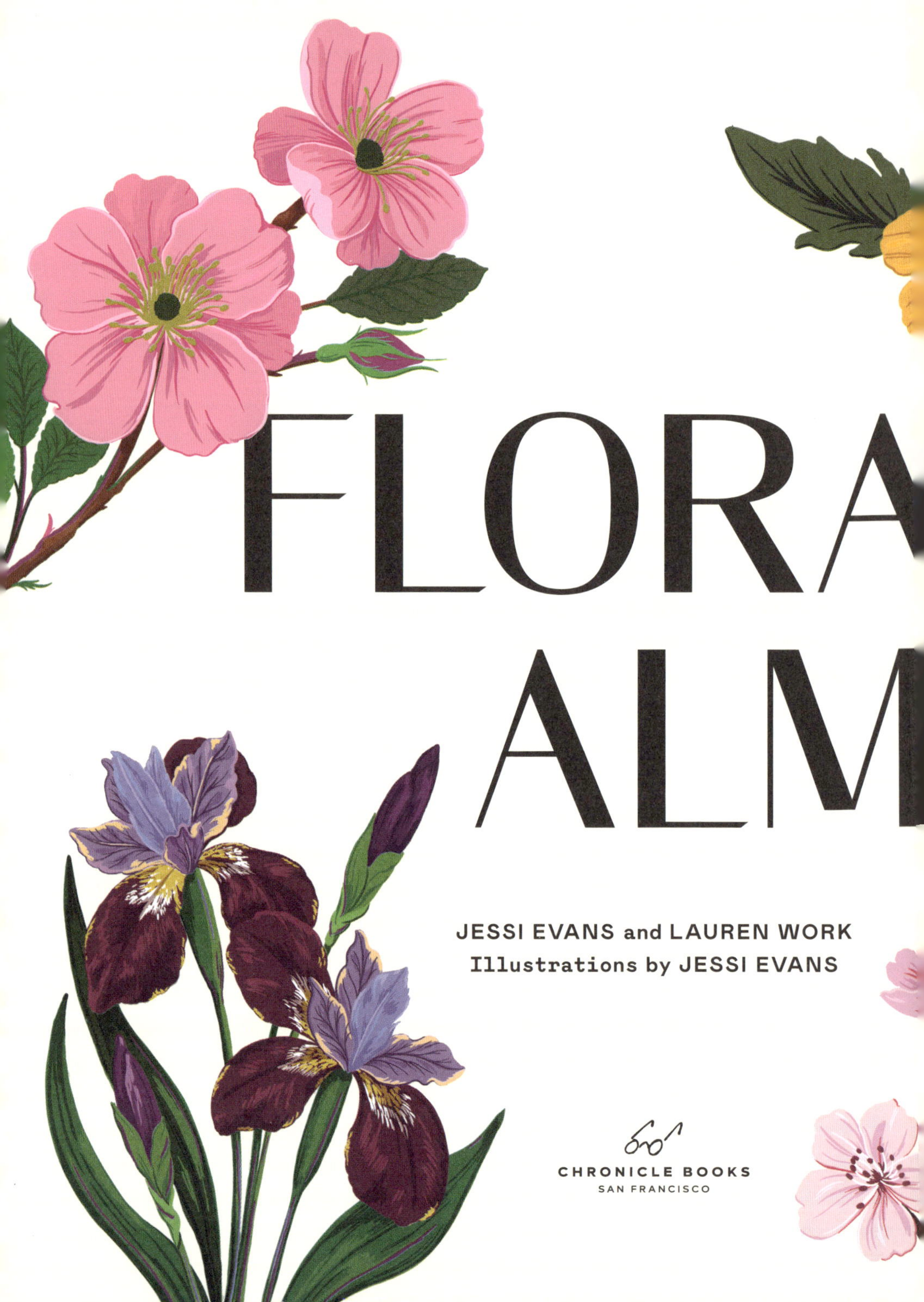

FLORA ALM

JESSI EVANS and LAUREN WORK
Illustrations by JESSI EVANS

CHRONICLE BOOKS
SAN FRANCISCO

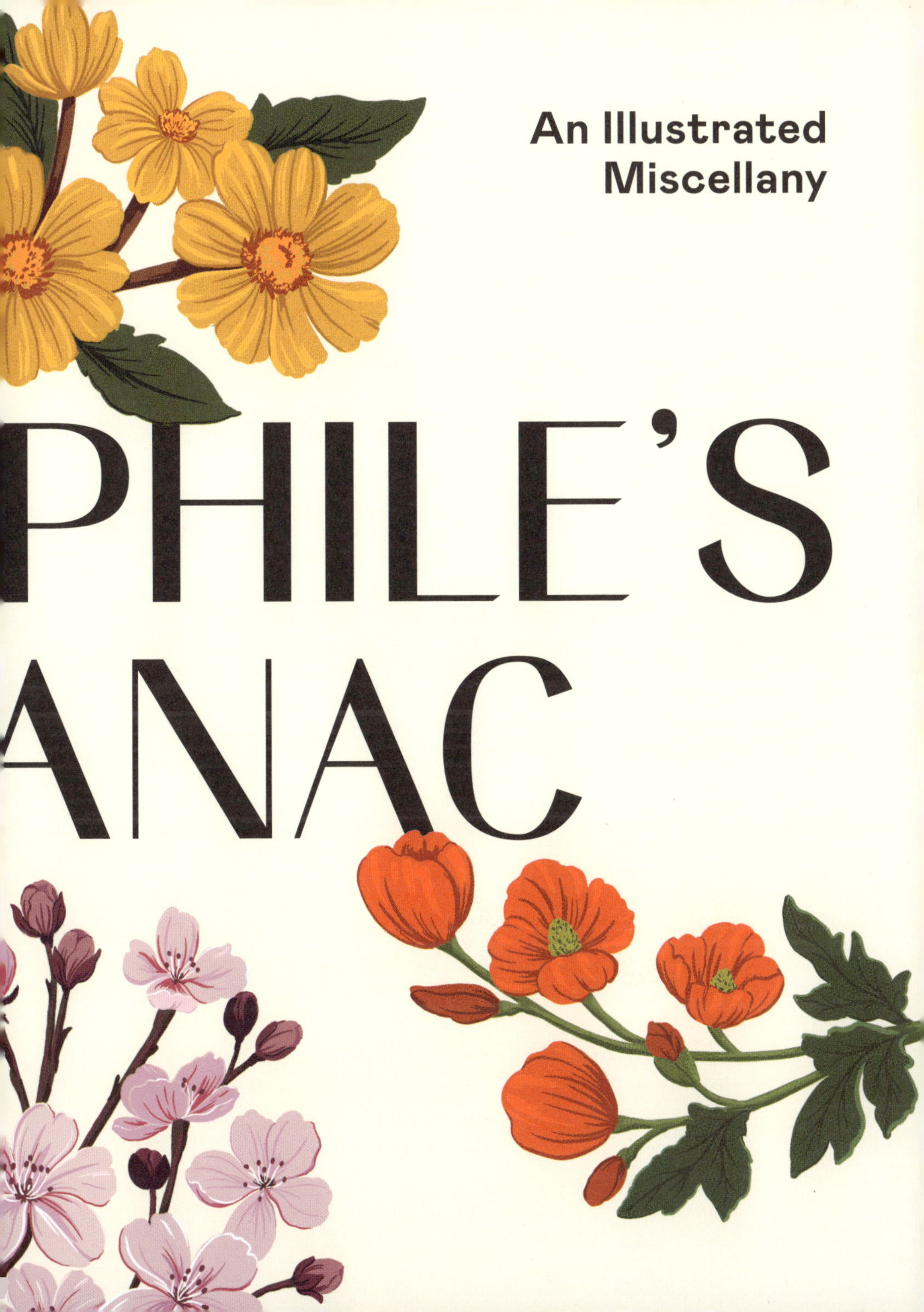
An Illustrated
Miscellany
PHILE'S
ANAC

Library of Congress Cataloging-in-Publication Data available.

ISBN 978-1-7972-3638-4

Manufactured in China.

Design by Wynne Au-Yeung.

10 9 8 7 6 5 4 3 2 1

Chronicle books and gifts are available at special quantity discounts to corporations, professional associations, literacy programs, and other organizations. For details and discount information, please contact our premiums department at corporatesales@chroniclebooks.com or at 1-800-759-0190.

Chronicle Books LLC
680 Second Street
San Francisco, California 94107
www.chroniclebooks.com

Part I

The Icons 10

Part II

The Blossoms 52

Part III

The Healers 88

CONTENTS

Part IV

The Wild Ones 132

Part V

The Oddities 188

INTRODUCTION

Our world is constantly and rapidly changing. News cycles spin, technology advances, the youth burgeon and become adults for a fleeting moment, until before we know it, the next generation of children grows up and inevitably takes their place. But some traditions are sacred, and the primal, ancient connection between flowers and humans is no exception.

We love to be in the presence of flowers. In fact, we bend over backward to make sure their vibrant color is always within reach: We plant them in our yards, in window boxes, and in community gardens. We cut them in bunches and put them in jars of water to bring their living beauty and fragrance indoors. We buy them at our favorite neighborhood florist, farmers' market, or grocery store and play florist for a day, trimming and combining them so they harmonize beautifully. We gift them to our loved ones at every milestone in life, from celebrations to farewells. We drop pennies into the vases to extend their time with us by even a few hours, forgetting the flower's greatest lesson: that everything is temporary. Flowers teach us to enjoy the present while it's available; that hoping to prevent the inevitable wilting is wishful thinking.

Think of this book as a love letter to flowers—their history, little quirks, defining qualities, folklore, and many uses. But more than that, it is also an exploration of the special bond we share with flowers and how they've been part of our rituals, remedies, art, and everyday lives for centuries.

Part I

The Icons

Some blooms need no introduction. More than flowers, these blossoms are poster children, textbook floral perfection. They are instantly recognizable, their legendary iconography passed down through generations. They are integral decor for life's most cherished moments—weddings, graduations, apologies, Valentine's Day celebrations, birthdays, just-because days. They have demanded our attention from their very first cultivation; they're the flowers etched into our deepest memories.

We remember the sweet perfume of roses in Grandma's garden, their scent lingering in our minds like a warm embrace. The first dahlias at the farmers' market remind us of the fleeting joys of late summer. Irises, gracefully emerging in late spring, herald rebirth and renewal. Fields of sunflowers, their suns on stems stretching toward the sky, speak to our nostalgia for childlike freedom and summer's endless possibilities. The vibrant poppy, its bold splash of red like a beating heart, reminds us that no matter where we roam, there's truly no place like home.

These flowers have storied histories, having journeyed across continents to thrive in the most unexpected places. They have flourished in royal gardens and wild meadows, in bustling cities and quiet byways, resilient against the test of time. Each flower in this chapter has won a hard-earned place here.

Rosa

Rose

Symbols & Meanings

What's in a name? That which we call a rose
By any other name would smell as sweet.
—William Shakespeare, *Romeo and Juliet*

Perhaps no other flower carries as much cultural weight as the rose. It is one of the most gifted flowers at celebrations, famously given for apologies and farewells, and in the hands of every couple come Valentine's Day. Some argue that the rose's overexposure detracts from its beauty, especially when considering the new-wave, contemporary florals that spurn common flowers to focus on rarer blooms. Though roses are common, they have remained timeless for good reason.

With their intricate beauty and intoxicating fragrance, roses are the epitome of romance. Their soft, delicate petals have captured hearts for centuries, symbolizing love, passion, and faithfulness. Beyond this, each variety holds a language all its own. Every color of rose conveys an intimate message, a heartfelt sentiment.

The number of roses you give carries its own language, as meaningful as their color. For instance, three roses whisper the timeless declaration of "I love you"; a baker's dozen suggests a secret admirer. Whatever color or amount you choose, each rose will communicate a heartfelt message to those you cherish.

Symbols & Meanings cont'd

RED ROSES

Bold and eye-catching, the red rose is the eternal symbol of passionate love and undying romance—a passion and devotion that will stand the test of time. To make a deep declaration of love, let the red rose do the talking.

PINK ROSES

A gentle, soft, and feminine expression of admiration and heartfelt gratitude, the pink rose is perfect for appreciating a loved one or proclaiming tender affection.

PEACH ROSES

The peach rose is perfect for showing your appreciation for someone or to honor a steadfast bond. The petals' warm, inviting glow symbolizes heartfelt gratitude and loyalty. A bouquet of peach roses shows those you care about how deeply you trust, respect, and care for them.

Symbols & Meanings cont'd

YELLOW ROSES

Radiating warmth and happiness, the yellow rose is a beacon of pure joy. Representing true friendship, these delightful petals capture the joy, laughter, and spark that a kindred spirit brings to your life.

WHITE ROSES

The symbolic pinnacle of purity, grace, and new beginnings, the white rose is a profound gift and marker of respect. These luminous beauties bring a sense of serenity and hope, making them perfect for moments of love, renewal, and new chapters.

PURPLE ROSES

Enchanting and elusive, the purple rose is among the rarest and most captivating of all the roses. Historically associated with royalty, these regal blooms symbolize love at first sight—the magic of instant, unspoken connection between individuals.

Folklore

The rose appears in countless cultures and timeless tales. Each story is unique and beautiful, attesting to the profound significance of roses as symbols of love, passion, and the enduring power of the heart. Their legacy is as rich and varied as the emotions they stir, a testament to how deeply and widely their beauty touches our lives.

In one version of a Greek myth, the rose was cherished by Aphrodite, the goddess of love. According to legend, upon learning of a plot to kill her mortal lover, Adonis, she raced to warn him, cutting her ankles on the thorns of roses. Her blood stained the white petals red, and by the time she reached Adonis, it was too late. He had been fatally wounded by a wild boar. Grief-stricken, she held him close as he took his last breath, her tears spilling among the stained roses. This legend has made its way around the world, creating a symbolic association between roses and love, loss, and devotion.

One version of Roman mythology tells of Cupid, the mischievous god of love, accidentally knocking over a bowl of wine with his wing as he stood beside Bacchus, the god of wine. The spilled wine magically turned into a rose bush, teaching us that love, like wine, can be intoxicating, unexpected, and all-encompassing.

History

Until recently, roses were believed to have originated in southern China around 500 to 400 BCE, although some evidence suggests they may be even older. According to recently discovered fossil records, roses were already flowering over thirty million years ago. The gorgeous roses we know today are primarily hybrids descended from those earliest varieties first seen in China, and they've been stealing hearts ever since.

The rose's allure has inspired countless stories throughout history, with two particularly significant moments leaving a lasting mark. In the seventeenth century, so cherished were these blooms, particularly their fragrant rose water, that they briefly accrued more value than gold. English royals treated them like treasure, using roses and rose water to barter for goods, settle debts, and even bribe for favors. Forget silver and gold—if you had a bouquet of roses, you were rich in the most beautiful currency of all.

Centuries later, red roses—by then known worldwide as the ultimate symbol of love—developed a surprising connection to horse racing. In the late 1800s, a delightful tradition was born: The Kentucky Derby winner was crowned with a garland of 554 stunning deep red roses. From that moment on, the race became known as the "Run for the Roses."

Botany

With more than thirty thousand captivating varieties, roses bloom in a stunning spectrum of colors. Revered for centuries for their elegance, fragrance, and the unique sweetness of their flavor, roses are a treasure cherished by cultures and hearts alike.

While most rose species call Asia home, a handful are native to North America, Europe, and northwest Africa. It was from a select group of fewer than ten species, primarily hailing from Asia, that the art of rose crossbreeding began. This delicate process, uniting two distinct varieties within the same species, gave birth to the vast array of garden roses we know and cherish today. Each rose variety is a beautiful testament to nature's creativity and the dedication of those who cultivate them.

Just like us, roses grow in different ways. They can shoot upright, climb gracefully, or spread as shrubs. Their rich, heavily scented petals are protected by the sharp thorns that arm the stems, accompanied by dark, lustrous green leaves with finely serrated edges. These two cutting features temper the sweetness of the flowers. The number of petals can vary dramatically, from as few as five petals to a lavish hundred. After the petals fade, the plant produces a berrylike fruit, the rose hip, which typically ranges from red to orange. Packed with vitamin C, rose hips can be transformed into delectable preserves or brewed into a soothing tea.

Recipe

ROSE PETAL ICE CUBES

Add a bit of rose-flavored magic to your chilled beverage. These beautiful cubes instantly elevate any drink, transforming it into a serene garden escape, perfect for a warm summer's day.

1 **To make, sprinkle a pinch of organic, pesticide-free dried rose petals and a touch of dried lavender into each compartment of an ice cube tray. Gently pour in filtered water and freeze for 3 to 4 hours.**

2 **Once they're frozen, pop out two or three cubes and drop them into your favorite beverage. Cheers to a refreshing, aromatic moment!**

`Note:` If you're feeling adventurous, try subbing in other liquids for water. Freeze cold brew with rose petals to add a dash of romance to your morning coffee, or enjoy a midday respite with green tea rose petal ice cubes.

Pop Culture

Since Queen Victoria's reign in the nineteenth century, it's become a royal tradition to have a flower named after prominent members of the royal family; often they are in the rose family. The tradition is still alive and well in the British monarchy, from the 'Princess Anne' rose to the clematis 'Meghan' (Markle, Duchess of Sussex).

Her Majesty the late Queen Elizabeth was a passionate flower lover and a regular attendee of the prestigious RHS Chelsea Flower Show, lending her name to many blooms over the years. But the crown jewel? The candy-pink 'Queen Elizabeth' rose, a stunning grandiflora created by rose breeder Dr. Walter Lammerts in honor of her 1953 coronation.

Red roses play a powerful role in *Alice in Wonderland*, and they're not there just to look pretty in the royal gardens. When the Queen of Hearts, infamous for her temper, enters the scene, even her loyal card soldiers are terrified. In a moment of panic, they realize they've planted white roses instead of red ones. To avoid a beheading, they rush to paint them red before the queen notices.

This moment isn't just haphazard chaos; it's rife with symbolism. White roses usually stand for purity and innocence, while red is the color of passion, danger, and aggression. So when those white roses are covered in red paint, it's like innocence is being smothered by fear and tyranny—perfectly mirroring the Queen's venomous temper.

Helianthus annuus

Sunflower

Symbols & Meanings

If ever a flower truly earned its radiant title, it must be the sunflower. With bright golden faces always turning toward the sun, they never miss a moment of its warmth and radiance. Their scientific name, *Helianthus*, derived from the Greek word for sun, *helios*, is a fitting moniker for a bloom that constantly reminds us to look toward the light. No wonder sunflowers symbolize loyalty and adoration, echoing their steadfast admiration of the sun. The sunflower serves as nature's own little love letter to us all.

Beyond their steady, faithful nature, sunflowers have a playful spirit! In the Sun tarot card, the sunflower shines as a symbol of positivity, freedom, and pure joy—a celebration of reconnecting with your inner spirit and a gentle reminder to let your heart be light and your spirit free.

Home

Few places are as deeply tied to the sunflower as the state of Kansas. The state motto, *Ad astra per aspera* ("To the stars through difficulties"), could aptly describe the sunflower's unbreakable spirit.

The state's extensive connection to the sunflower is no accident. The Sunflower State provides the perfect growing conditions for this flower to flourish.

But for Kansans, the sunflower is more than just a pretty bloom; it is part of the fabric of their history. It tells the story of struggle and triumph, of winding trails and boundless prairies, of the frontier spirit that shaped the state. The sunflower stands tall as a tribute to the past, a reflection of present pride, and a promise of a golden future.

XIX
THE SUN

Folklore

The sunflower's story is one of devotion even in the face of heartbreak. In Greek mythology, the nymph Clytie fell hopelessly in love with Apollo, the radiant god of the sun. For a while, Apollo returned her affection, but soon his gaze wandered to another.

Overcome with jealousy, Clytie swore revenge on this competing nymph, which drove Apollo into a fierce fury. In his anger he transformed Clytie into a sunflower, forever immortalized in golden petals. Yet even in her new form, her love never wavered. She turned her golden face toward Apollo, following his journey across the sky each day—just as sunflowers faithfully trace the path of the sun.

History

These joy-filled beauties are estimated to have been formally cultivated for the first time between 3000 BCE to 1000 BCE in the Americas, where they took root in the lands we now call Arizona and New Mexico. Early Indigenous civilizations cherished them not only for their beauty but also for their many uses. The nourishing seeds were ground into flour to bake bread, while their vitamin-packed oil was perfect for moisturizing skin and hair. Even the sturdy stalks of the sunflower provided ample building materials. Some evidence suggests that sunflowers were cultivated even before corn, underscoring their deep-rooted significance for ancient civilizations.

When Spanish explorers introduced the sunflower to Europe in the 1500s, it quickly became a cherished ornamental bloom, grown as a showpiece in many gardens. Beyond its beauty, the sunflower has medicinal properties, with seeds used to manage pulmonary ailments such as coughs and the common cold.

The sunflower's journey took an unexpected turn as it gained popularity in Russia, where its oil became highly desired. The Russian Orthodox church aided its growing popularity by giving exemptions to the sunflower—while most seed oils were prohibited during Lent, sunflower oil was allowed. The surge in demand for sunflower oil led to the development of hybridized varieties of the flower, which eventually made their way around the world.

Craft

LET'S GROW SUNFLOWERS!

Growing sunflowers is a simple and rewarding process. Follow these steps to help these radiant blooms thrive in your garden:

1. `Pick the right location:` Sunflowers love and need full sun, so select a location that receives at least 6 to 8 hours of sunlight per day.

2. `Prepare the soil:` Sunflowers can survive in most soils but prefer well-drained, slightly acidic to neutral soil. Add compost or organic matter to improve soil structure and drainage.

3. `Plant the seeds:` Sow sunflower seeds outdoors in late spring, after the last frost, when the soil temperature is consistently above 50°F [10°C]. Sow seeds 1 to 2 in [2.5 to 5 cm] deep and water gently. Keep the soil consistently moist until seedlings emerge. Each variety may have specific requirements, so be sure to check your seed packet for details.

4. If you want to give the flowers a head start, sow seeds indoors 2 to 3 weeks before the last frost. Transplant the seedlings when all frost danger has passed and they have two sets of true leaves, being careful not to disturb the roots.

5 `Care for the seedlings:` Sunflowers are quite drought tolerant but will be happiest with regular watering. Tackling weeds early on will remove competition and lead to healthier plants.

6 Time to enjoy your sunflowers! You can either leave them to provide food for the wildlife in your garden or bring them into your home as beautiful cut flowers. Cut in the early morning when the flowers are hydrated; choose flowers with petals only partially open to ensure a long vase life.

Botany

Sunflowers are not only beautiful to look at but also hardy and resilient, considering their remarkably short growing season. On average, it takes just ninety days for a sunflower to grow from seed to full bloom.

With around seventy different species, their diversity is impressive. The compact and fluffy 'Teddy Bear' sunflower reaches just around 3 ft [90 cm], while the towering 'American Giant' can soar up to 16 ft [4.9 m].

It's not just their size that varies—sunflowers come in an array of colors. While the classic golden petals with deep-brown centers are best known and most common, look for unexpected varieties like 'Chocolate', with velvety burgundy-brown petals kissed with hints of gold, or 'Strawberry Blonde', whose ruby-red petals fade into soft pinks with buttery yellow tips.

Botany

cont'd

Sunflowers not only provide nourishment for both humans and wildlife but also play a crucial role in healing the earth. They are hyperaccumulators, meaning they can absorb toxins from the soil. After the devastating nuclear disasters at Hiroshima, Fukushima, and Chernobyl, vast fields of sunflowers were planted to extract harmful metals and radioactivity from the land. With each golden bloom, they work to restore balance—proof that nature, when given the chance, will always find a way to heal.

Pollinators

When you add sunflowers to your garden, you will find they attract pollinating insects, such as swallowtail butterflies, drawn to the plant's sweet, warm nectar. Found across much of the United States, males are adorned with bold rows of yellow spots; females have delicate markings and a pop of sapphire blue.

Bees also play a huge role in helping sunflowers grow seeds, even though sunflowers in the United States were specially bred to fertilize themselves. Both native bees and honey bees boost pollination, making seed production way more successful.

Some native bees are so intertwined with sunflowers, scientists simply call them "sunflower bees." Belonging to the Megachilidae family, sunflower bees are solitary, meaning they do not belong to a hive. They are not aggressive and play an important but peaceful role in pollinating sunflowers.

Dahlia

Symbols & Meanings

Bold, dynamic, and deeply symbolic, dahlias take center stage no matter where they bloom. These striking beauties represent elegance, creativity, and growth. Their ability to flourish in tough conditions is a testament to their resilience, making them living emblems of inner strength.

Dahlias also represent the power of transformation. A dahlia, born from a humble tuber, is destined to produce an abundance of radiant blooms. The journey is slow and steady, the transformation gradual and profound. From the dahlia's quiet and dark beginning, it emerges and basks in the sun, proving that with patience and the right conditions, beauty can reach its fullest potential.

In the Victorian era, dahlias symbolized an eternal bond, representing a love that will stand the test of time. Whether a symbol of resilience, transformation, or enduring love, the dahlia reminds us that even in the face of change, true beauty and strength will always find a way to flourish.

Home

The dahlia's ancestral home is in the mountains of Mexico and Central America, where it has thrived for centuries. Ranging from deserts to forests, these regions teem with rich biodiversity, well suited to a flowering plant like the dahlia.

When sixteenth-century Spanish conquistadors ventured into the Aztec empire, their mission was not only to conquer but also to explore. Among the native treasures they claimed for themselves was the towering bloom now known as the tree dahlia (*Dahlia imperialis*). This colonial acquisition marked the beginning of the dahlia's journey to Europe.

The dahlia, officially designated as the national flower of Mexico in 1963, is called *atlcocotlixochitl* or *acocóxitl* ("flower of the water stems" or "water pipe flowers") in the native Nahuatl language. The name derives from Aztec use of hollow dahlia stems to transport water.

The Mexican Dahlia Association focuses on dahlias in rural areas of the country and has become a vital force in both conservation and appreciation of this remarkable bloom. Through the association's efforts, various species are grown not only for their statement blooms but also for their nutrient-dense qualities. These are incorporated into a variety of edible products, from syrup to cookies, showcasing the versatility of the plant used in Mexican culture to this day.

History

In 1791, shortly after the flower was brought to Europe by Spanish colonizers, the dahlia received its official botanical name. Antonio José Cavanilles, director of the Royal Gardens of Madrid, chose the name "dahlia" to honor the Swedish naturalist Anders Dahl.

In the eighteenth century, Spain controlled commerce in the valuable red dye cochineal, made from insects native to Mexico. Tired of depending on Spain, the French hatched a plan to smuggle these insects into France. By sheer accident, the shipment also included dahlia tubers. Although the insects perished, the dahlias thrived.

Seeing an opportunity, the French minister offered these newfound flowers to Empress Josephine Bonaparte, a talented gardener with a keen eye for beautiful plants. In her care, the dahlias thrived and became the envy of many. She was very protective of her plants, so she was shocked and furious upon learning that a visiting Polish count had successfully stolen over a hundred of them. Her response? Destroy all remaining tubers! Josephine died at fifty-one, but her stolen dahlias spread across Europe, becoming widely popular in Germany and ultimately immortalizing her as a dahlia enthusiast.

Today, dahlias are a dazzling showpiece in gardens around the world. Originally there were only forty-two species, but their popularity has sparked the creation of over sixty thousand cultivated varieties registered with the International Dahlia Register.

Botany

The dahlia is second only to the tulip as the world's most cultivated flower. They are prized for their continuous blooms from midsummer into fall, wherever they are grown.

The dahlia has a diverse range of sizes and forms that work beautifully in cut flower arrangements. For instance, a dinner plate dahlia plant can stretch upward of 6 ft [182 cm] and produce flowers as large as 12 in [30 cm] in diameter. The fifty-seven thousand dahlia varieties are organized in fourteen different groups based on their flower shape.

Festival

Swan Island Dahlias in Canby, Oregon, is the largest and leading US dahlia grower. Each August and September they host their free Annual Dahlia Festival, a cherished and celebrated tradition. The display of over 375 varieties in an array of beautiful colors draws dahlia lovers from near and far.

Guests can stroll through nearly fifty acres of stunning gardens with dahlias in full bloom. Whether you learn the secrets of dahlia care, try your hand at floral design, or simply drink in one of nature's most exquisite and varied blooms, something special awaits at every turn.

THE SWAN ISLAND DAHLIAS FARM
Dahlia
Festival
AUGUST THROUGH SEPTEMBER

Iris

Symbols & Meanings

In ancient Egypt, the iris was seen as more than just a flower. It was an emblem of rebirth and eternal life. Its three delicate petals were believed to hold the keys to the trust, knowledge, and courage required for a soul's journey into the afterlife.

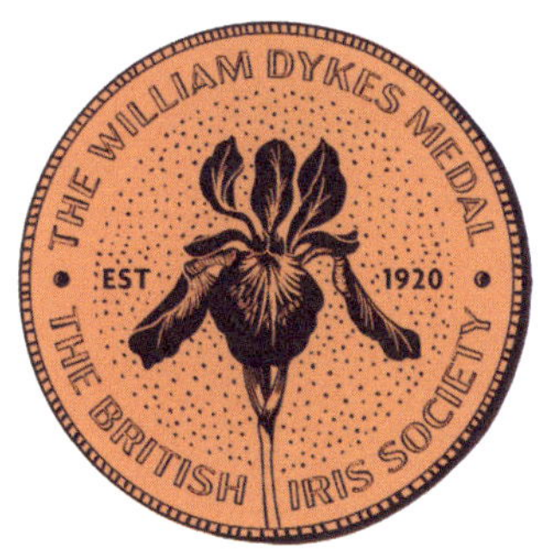

History

For millennia, the iris has graced gardens with its regal beauty. Its first appearance dates back to 1469 BCE, when King Thutmose III of Egypt introduced the bloom to his royal gardens. A passionate gardener, Thutmose had discovered the iris during his conquest of Syria—and was captivated. As their allure spread, irises appeared throughout Egypt, adorning temples and tombs. By the 1600s, the iris had reached the colonial United States.

In the nineteenth century, physician and professor Michael Foster became the first dedicated iris enthusiast in the United States. He was devoted to his garden and the study of irises, collecting almost every then-known variety and hybridizing them as he learned more.

Foster entrusted his legacy of work to his friend, horticulturist Ellen Willmott. She in turn shared his notes with an eager botanist, William Rickatson Dykes. Dykes began a breeding program and published *The Genus Iris*, which became the primary resource for enthusiasts everywhere. Dykes was recognized for his contributions to the iris world; today the highest honor that can be bestowed upon an iris is the Dykes Medal. This coveted award has ensured his legacy along with the continued enthusiasm for irises.

Botany

The American Iris Society classifies three main types of irises: aril, bearded, and beardless. Each has its own charm, with a mesmerizing range of colors, shapes, and blooming styles. While the imported irises are stunning, there are twenty-eight equally arresting varieties native to the United States, found in woodlands, fields, and marshes.

Beyond these beloved varieties, the genus *Iris*, a captivating member of the Iridaceae family, includes around three hundred delightful species. Most are perennials, emerging from rhizomes to reveal bold blossoms. While most varieties flourish in cooler temperatures, some of the most breathtaking species thrive in the sun-drenched regions of the Mediterranean and Central Asia.

For centuries, iris roots have been cherished for their medicinal properties, offering remedies for everything from skin infections to syphilis. Today they are still used in alternative medicine, purported to support liver health and purification.

The air-dried roots take on a delicate scent, reminiscent of a violet, that deepens over time. Although once used in several beauty products, the roots were later discovered to cause skin reactions, so use shifted toward aromatherapy blends and scented sachets.

Art

Vincent van Gogh's masterpiece *Irises* expresses his admiration of the flower. He meticulously studied the delicate shapes and vibrant colors, capturing the iris's graceful twists and turns. Created during an asylum stay, *Irises* suggests that despite his inner turmoil, he found joy and serenity in these blooms.

Another artist enamored with the iris was Georgia O'Keeffe. In *Light Iris* and *Black Iris* she magnified the flower, transforming it into abstract, sculptural compositions famously suggestive of female anatomy. Yet O'Keeffe herself rejected these assumptions. Believing the beauty of flowers went unnoticed in the rush of everyday life, she wanted to draw attention to their intricate details. By enlarging the blooms to monumental proportions, she created works that demanded attention. O'Keeffe reminds us how intimate and inspiring nature can be.

Folklore

The name "iris" comes from *iris*, the Greek word for "rainbow," and Iris, the goddess of rainbows. Iris was a cherished messenger of the mighty Hera, queen of the gods. It is said that when the gods desired contact with a mortal, they sent Iris in their stead, her golden wings leaving a brilliant rainbow in her wake.

The ancient Greeks planted the purple iris on women's graves, a tender offering meant to call upon the goddess Iris herself. It was believed she would lead the souls of the departed safely on their journey to the afterlife. This heartfelt tradition lives on in modern-day Greece.

Papaver

Poppy

Symbols & Meanings

The wistful poppy. Used to pay tribute to those who've passed on to the afterlife, this flower is a beacon of hope, honor, and remembrance.

The poppy's status as a symbol of remembrance bloomed out of the darkness of the First World War. This enduring symbol of hope grew on landscapes scarred by war, where, amid sorrow and violence, the poppies persisted, a silent yet powerful embodiment of strength. The poppy thus became associated with Memorial Day in the United States.

Home

California's reputation as a warm, sunny region was secured in part by the dazzling California poppy, *Eschscholzia californica*—the state flower as of 1903.

From late spring to early summer, these cheerful blooms come forth, blanketing hills and valleys across California in a sea of brilliant flowers. Each year on April 6, California celebrates Poppy Day.

Pop Culture

The opium poppy, *Papaver somniferum*, occupies a mysterious place in floral history that even the silver screen couldn't resist. In *The Wizard of Oz,* the fragrant waves emanating from a field of poppies famously cast a spell over Dorothy, the Lion, and Toto, lulling them into a deep slumber.

The opium poppy carries a legacy of both healing and darkness as one of the world's oldest botanical remedies. It's revered for its ability to relieve everything from physical pain to sleepless nights—but this potent effect comes at a cost. In L. Frank Baum's tale, the poppy's enchantment mirrors the flower's complex nature: What can lull us into a peaceful rest can also cast a dangerous, addictive net. Baum was making a cultural commentary on opium use, which had reached a peak in the United States.

Folklore

Somnus, the Roman god of sleep, is often portrayed either in a cave encircled by poppies or drifting into slumber, cradling a bouquet of poppies. From this godly connection, the association of poppies and peaceful slumber was forged.

Another Roman deity closely tied to the poppy is Ceres, the goddess of harvest. In ancient depictions she holds a wheat sheaf and a poppy in either hand, as they were often grown together. The poppy remains a beautiful reminder that while hard work is essential for growth and abundance, so too is the rest that allows us to recover and replenish.

History

The poppy dates back to 3400 BC in the Middle East. Initially, the poppy was revered not for its beauty but for its medicinal and recreational properties as a relaxant and sedative.

As knowledge of the poppy's potent properties spread, so too did demand for it, leading the humble flower to the Silk Road, a network of trade routes that connected the Mediterranean to Asia. Along these routes the opium trade flourished and eventually led to the Opium Wars of the nineteenth century, a trading dispute between Great Britain and China that developed into a full-scale naval war—one of the most significant conflicts in history to be caused by a plant.

Today, the poppy remains the primary source for extracting morphine and codeine, two potent medications that have become essential in modern medicine for pain relief.

Pollinators

Bees and poppies share a symbiotic relationship that is essential for the reproduction of poppies and the survival of bees. The bees are drawn to the poppy's bright blooms and large, open petals, which make it easy to reach the nectar. The Oriental poppy (*Papaver orientale*) is a top-choice bloom for our buzzing friends, giving them sustenance during a time when other plants decline in pollen and nectar. The poppy mason bee (*Hoplitis papaveris*) is a unique little bee that creates its nest within the petals of the common poppy.

Save the bees by adding poppies to your home garden! Not only are there endless varieties and colors, but these special flowers, like others, can increase the health of the entire ecosystem.

zuzu's
petals
it's a wonderful life
special
Roses
BROO

FLOWER SHOP

Zuzu's Petals

Brooklyn, New York City | Blooming since 1971

For over fifty years, Zuzu's Petals has been a landmark in the Park Slope neighborhood of Brooklyn, making an impression on both the ordinary and extraordinary moments in their customers' lives. Upon entering through a white picket fence, prepare to be greeted by a jolly golden retriever and an abundance of perfectly cut fresh flowers, the feeling of spring bottled up in a brick-and-mortar.

To trade in flowers is a noble and rewarding occupation—despite the brevity of the transaction, it's far from inconsequential. Working among vases of roses, delphiniums, and freesias, owner Fonda Sara connects with folks while they choose their flowers, sharing the intricacies of their lives. Over time and through repetition, they form personal bonds and precious relationships.

Everything chosen for Zuzu's Petals reflects Fonda's passion for living things. The shop's namesake is Zuzu, George Bailey's youngest daughter in the movie *It's a Wonderful Life*. Fonda's life with her shop has been just that.

WHY FLOWERS?

A life with flowers. A long time ago, someone told me that if you do something you love, you'll never work a day in your life. After fifty-odd years running a flower shop, that isn't always the case, but it has been true often enough to keep me passionate for things that grow. A bundle of deep chocolate ranunculus or fat peachy pink peonies will always fill me with indescribable joy. I share that joy with everyone who has passed through the front door of my shop and implanted it in the hearts of many people who have worked for me.

—Fonda Sara, owner, Zuzu's Petals

Part II

The Blossoms

A common misconception is that winter is a time of rest for all living things. For humans and animals, yes, but for the blooms featured in this chapter, it is a time of work, all happening silently but diligently beneath your feet.

After the long dark chill, just when you think you can't take one more second of gray skies, a breathtaking floral spectacle unfolds and rescues you from the winter blues. Whether a large magnolia bloom, a dramatic wisteria cascade, the perfect pink petal of a cherry blossom, the sophisticated shape of a camellia, or the absolute revelation of a flowering dogwood, these blossoms are the very picture of spring.

And as these flowers emerge, they bring our favorite creatures out of winter hibernation, bursting into activity after months of solitude and quiet. Squirrels, birds, bees, butterflies—all seek refuge in the plants' billowing petals, leaves, or stems.

Magnolia

Magnolia

Symbols & Meanings

For well over a million years the magnolia tree has been a strong, steady presence on our planet. Wise, resilient, and enduring, magnolias are hardy and have stood the test of time, hence their frequent appearance in wedding bouquets despite their brief flowering season.

Depending on the variety, magnolias can grow up to eighty feet tall, towering over their subjects in the plant kingdom, embodying dignity and stature.

Home

The magnolia holds the honor of being the state flower chosen in two states—Louisiana and Mississippi. However, Mississippi's love of the magnolia flower is unrivaled: Not only is it the state flower, but it is also the state tree *and* appears on the state flag. Mississippi school children were asked to vote for their favorite flower, and the magnolia was crowned the winner in a landslide.

Pop Culture

This versatile blossoming tree has made it onto the big screen many times, but perhaps its most iconic and memorable appearance was in *Steel Magnolias*.

Originally written as a stage play by Robert Harding to honor the women who raised him and weathered hardships, *Steel Magnolias* is a beautiful exploration of the resilience of women and their relationships, connection, and love found, lost, and found again. The movie seems to say that although women are perceived as fragile and delicate as a magnolia petal, they are forged like the hardest of metals and can persist through troubled times.

Botany

Magnolia trees are one of the oldest flowering plants. Fossil records show that magnolias thrived alongside the dinosaurs one hundred million years ago in what is now North America, Asia, and Europe.

Today, there are an estimated 240 documented magnolia species. They can be evergreen or deciduous, large trees or shrubs.

Pollinators

Given the magnolia's hundred-million-year history, its pollination differs from that of more recent plant families. Since bees did not yet exist in the ancient environment, beetles became the magnolia's primary pollinator.

To help these small-but-mighty beetles thrive, the magnolia produces massive amounts of pollen, which the beetles consume as fuel. Thus a beautiful and symbiotic relationship between the magnolia and the beetle was formed and has stood the test of time.

Craft

SOUTHERN MAGNOLIA WREATH

Embrace Southern hospitality and make a magnolia wreath to welcome guests into your home with warmth, charm, and grace. At florist shops, you'll find magnolia branches or "tips" that can be cut into single stems.

20 magnolia tips

Clippers

12 in [30 cm] wreath form

Paddle wire

Wire cutters

1. **Begin by cutting the magnolia tips into 6 in [15 cm] pieces. Gather 2 or 3 pieces together to form 8 to 10 small bundles.**
2. **Layer the bundles to overlap one another, arranging them on the wreath form to create a lush, full appearance. As you place each bundle, secure the stems to the wreath form using the paddle wire, ensuring that each piece stays in place.**
3. **Continue layering and wiring the magnolia bundles until the full circle of the wreath is complete.**
4. **Hang your stunning wreath for all to see! If stored properly, this wreath will dry and preserve well and can welcome visitors for years to come.**

Cerasus serrulata

Cherry Blossom

Symbols & Meanings

The cherry blossom (*sakura* in Japanese) is inextricably linked to Japanese history, culture, traditions, and mythology. It was widely believed that a harvest deity lived within the cherry tree, so farmers would pray and offer gifts to the tree, enticing the deity to yield a fruitful harvest.

Each spring, an intoxicating sea of pink and white flutters across the Japanese countryside and cities. The cherry blossom's bloom time is all the sweeter for being so brief. Its short yet impactful life represents the impermanence of all things and the delicate balance of beginnings and endings.

In Greek mythology, it was believed that the cherry blossom contained a potion that would grant the gods immortality. Similarly, in Chinese folklore, the phoenix rests on a bed of cherry blossoms to gain everlasting life.

The cherry blossom represents bounty, the fleeting nature of life, and the impossible desire to make it last forever. A multifaceted flower indeed.

Botany

Like almonds, peaches, and apricots, the cherry blossom belongs to the genus *Prunus*. The lifespan of the cherry tree is normally fifteen to thirty years. Native to Asia, they are grown worldwide in similar climates. In North America, the cherry blossoms of Washington, DC, and Seattle, Washington, are renowned.

The deciduous cherry tree sheds its leaves annually. The branches spread out in a wide canopy, creating an expansive bloom display in early spring. The vibrant foliage is just as captivating as the blossoms, whether they're the classic early-season red, mid-season green, or postseason autumnal brown and orange. Unlike the cherry species cultivated in orchards for their fruit, the ornamental varieties are bred for their blossoms; a few species produce very small, inedible fruit.

Art

Cherry blossom motifs are seen in many Japanese art forms, from paintings and woodblock prints to pottery. The classic pink-and-white blossoms and sinuous branches have been featured in the work of Japanese artists such as Utagawa Hiroshige, Katsushika Hokusai, and Mori Togaku.

The West was introduced to the cherry blossom when trade renewed with Japan in the mid-1800s. The burgeoning art movement and the increased flow of goods from Japan had Western artists clamoring over the woodblock prints. Claude Monet, Edgar Degas, and Vincent van Gogh were among the artists inspired by Japanese cherry blossom artwork.

Folklore

Japanese culture offers the beautifully selfless folkloric tale of *Uba-zakura*—"old woman cherry" or "milk nurse cherry tree"—in which a wet nurse gives her life and soul to spare the life of a child in her care. Today, some believe that cherry blossoms bloom on the anniversary of her death.

Festival

A generous gift of cherry trees from Japan in 1912 sparked the beautiful tradition of the annual National Cherry Blossom Festival in Washington, DC. The festival celebrates the spectacular display for around four weeks, with events and activities honoring American and Japanese cultures.

Mother Nature can be unpredictable and fickle, so though the tentative blooming date is April 4, National Park Service horticulturists can make an accurate prediction only within ten days of blossoming. So every year the festival dates are scheduled in advance, and cherry blossom lovers hope for the best!

まつり
JAPAN

Wisteria

Symbols & Meanings

After a long dark winter, few spring sights are more welcome than the breathtaking purple-hued canopy formed by wisterias in full bloom. Wisterias are beloved around the world, giving rise to a myriad of different symbolic meanings.

In the Victorian era, the wisteria represented clinging tightly to romantic love. It's easy to see how the wisteria vine's need to hold steadfast and true to its earthly bonds mirrors love's enduring nature. The story goes that if a couple sits under the protective shelter of a wisteria trellis, they will fall in love so deeply that it will last for eternity.

The ancient Chinese practice of feng shui focuses on harmony between an individual and their surroundings. According to feng shui, incorporating wisteria into your space invites a sense of humility, the gentle swaying of its dangling flowers representing a reverent, peaceful bow.

Recognizing wisteria's impressive lifespan (the plant can flourish and endure for over one thousand years!), many cultures associate the flower with immortality and longevity. In fact, there is a twelve-hundred-year-old wisteria thriving in Kasukabe, Japan.

Botany

Chinese wisteria is a perennial bloom that will brighten your garden and life for many years to come, though you should plant with caution as it's an invasive species. With blooms hanging down like the branches of a weeping willow tree, these flowers are known to attract butterflies and pixies alike, filling your garden with springtime creatures. In May, when the leaves emerge, wisteria bursts into bloom in gorgeous clusters of purples and whites, with a deliciously light fragrance that announces: "Summer is on its way."

The wisteria, as a member of the pea family Fabaceae, is actually a legume! Legumes have long been regarded as valuable soil-building crops, improving soil's biological, chemical, and physical properties. Wisteria not only enriches the soil but also gives companion plants a nutritional boost.

The wisteria's flowers are a delight for all the senses. The blossoms can be gently transformed into a sweet, fragrant jam, while the leaves brew into a subtle, slightly bitter tea. But beware: Though the blossoms are edible, the rest of the plant is toxic to consume. Always forage with care.

History

Originating in China, Japan, and parts of North America, this fast-growing vine was valued for its practical applications. As its popularity grew, the Japanese took note of its potential as a garden plant. Collectors brought wisteria to Europe in the 1800s.

Though unconfirmed, it's suspected that the English botanist and zoologist Thomas Nuttall named the wisteria after the Wister family in Philadelphia, with whom he worked closely and shared a passion for botany.

Folklore

Italian folklore tells of a young shepherdess weeping over her perceived lack of beauty. Her body curled and twisted around a tree trunk, and her sorrowful tears were transformed into delicate clusters of flowers.

Festival

The Sierra Madre Wistaria Festival celebrates connection, community, and nature. You'll find a joyful collective focused on supporting small businesses, boosting the local economy, and making a better life for residents of Sierra Madre, California. At the heart of the festival is a breathtaking wisteria, the world's largest blooming plant, recognized by Guinness World Records!

Pollinators

The wisteria's most elegant and popular pollinator is the hummingbird. With their iridescent wings and lightning-fast movements, these tiny acrobats delicately gather nectar and inadvertently transfer pollen from bloom to bloom. Though hummingbirds are said to prefer red, orange, yellow, and pink flowers, the wisteria lures them with sweet scent and ample nectar.

Camellia

Camellia

Symbols & Meanings

The camellias most often grown in gardens (*Camellia japonica*, *Camellia sasanqua*), like the somewhat similar rose, come in various shades, each with their own particular symbolism.

WHITE CAMELLIAS

The white camellia evokes the unconditional love that binds family and friends together, especially in times of deep loss.

PINK CAMELLIAS

The soft, tender nature of the pink camellia represents the longing and physical ache of missing someone near and dear to your heart.

RED CAMELLIAS

The red camellia represents passion, desire, and the heart's most fervent yearnings.

Home

For over two centuries Alabamians have chosen the camellia to adorn their clothing, grace their gardens, and elevate their tablescapes. In 1959, thanks to the discerning eye of Governor John Patterson's wife, Mary Jo, the camellia was crowned Alabama's official state flower.

Pop Culture

Coco Chanel was haute couture personified, and when she pinned a delicate white silk camellia to her belt in 1913, Chanel cemented its place in the fashion hall of fame. Completely mesmerized by the flower's simple elegance, Chanel made the camellia a fashion staple by incorporating it into everything she wore and designed, from embroidered dresses to jewelry and more. The flower became her signature, forever associated with the timeless sophistication of the Chanel brand.

History

The camellia originated in lush landscapes of eastern and Southeast Asia, from the Himalayan regions of China to Japan and Indonesia.

By the mid-1700s, these alluring blooms had captivated the Western world, quickly becoming a symbol of grace, elegance, and rarity, especially during the opulent Victorian era. They ascended rapidly in high society, with only the wealthiest families cultivating them in private conservatories.

For over a century, the camellia was reserved for the most exclusive gardens, admired for its refined beauty and delicate, layered petals. However, as gardeners discovered the plant's resilience, it began to flourish on a larger scale, leading to a thriving cut flower industry.

Botany

Camellias bloom in late winter, gracing gardens and landscapes with vibrant color when little else is blooming. In addition to showy flowers, their glossy evergreen leaves stay stunning all year long.

Camellias are a bit more finicky than the other flowers in this chapter: They require the good stuff—well-drained, moisture-retentive, acidic soil rich in organic matter—and they thrive in partial shade. The blooms can span up to 5 to 6 in [12 to 15 cm], depending on the type. If you're looking to add a touch of love to your garden, you can't go wrong with a camellia—it's a flower that keeps on giving, year after year.

Festival

Tuscany in springtime offers a breathtaking experience, the Mostra delle Antiche Camelia della Lucchesia festival. During a three-week period in March, the camellia gardens in and around the historic city of Lucca, Italy, are opened to the public, offering a view of the ancient architecture and Tuscan landscape as spring breathes new life into the region.

Introduced to Italy in the late eighteenth century, the camellia thrived. The temperate Tuscan climate is so pleasing to camellias that in the early aughts, the International Camellia Society planted the first camellias on the forested slopes of a nearby peak, Monte Serra, which today counts over forty thousand camellia plants.

The Mostra delle Antiche Camelia della Lucchesia festival offers an opportunity to appreciate the connections between nature, art, and culture from centuries past to modern times, a unique experience for those interested in Tuscany's botanical heritage.

Recipe

CAMELLIA TEA

Camellia sinensis leaves are fermented in stages to create many beloved teas, from delicate whites to vibrant greens, rich pu'ers, fragrant oolongs, and bold blacks.

Indulge in the soothing pleasure of green tea made with camellia leaves. Here's your guide to crafting the perfect cup:

1 **Before you begin brewing, warm your favorite teacup or mug. Pour in a bit of hot water, swirl it around, then pour it out.**

2 **Add 2 cups [16 oz] cold water to a teakettle and bring it to a gentle simmer, about 180 to 205°F [82 to 96°C]. Let it rest for just a moment, so the temperature isn't too hot for the delicate green tea leaves, about 145 to 170°F [63 to 77°C].**

3 **Place loose green tea in a tea strainer, about 1 teaspoon of loose tea for every 6 oz [180 ml] cup. Pour in the hot water and allow the tea to steep for 2 minutes. If you like a touch of sweetness, add a drizzle of maple syrup or honey.**

Cornus

Dogwood

Symbols & Meanings

The dogwood is a resilient plant, with its sprawling boughs and distinctive blossoms standing as a beautiful reminder that life can not only survive but thrive, even after the harshest of winters. Its distinctive blooms signal not just the end of winter but the beginning of something new—a symbol of rebirth and renewal.

Rebirth and renewal bring us to the legend of the dogwood. According to this legend, the dogwood tree once played a major role in the crucifixion of Jesus. Its four large bracts—those outer leaves that look like petals—are thought to symbolize the shape of the cross, the four nail wounds, and even the crown of thorns. It's not a historical fact but rather a poetic tale. Whether you're a believer or just someone who appreciates a good story, the legend of the dogwood offers a beautiful reflection on hope and transformation.

The dogwood has long symbolized love and affection, especially during the Victorian era when each flower spoke its own language. Back then, a hopeful bachelor might offer a sprig of dogwood to the one who'd captured his affection. If she returned the bloom, his hopes would wither, for that meant his feelings were not reciprocated. But if she chose to keep the flower, it was a silent promise that she returned his affection, and their shared love could unfold.

Folklore

Cherokee legend tells of the Dogwood People who lived in the heart of dogwood groves. So small they were scarcely visible, these spirits were sent to teach humankind the art of living in harmony with nature.

Kindhearted and compassionate, these caretakers of the vulnerable and innocent tended to babies and looked after the elderly, embodying a quiet love for all life. Their presence was felt in the rustling leaves and the wind's soft whispers, reminding all who wandered near the dogwood trees to be gentle and kind and to practice reverence for all living things.

History

The flowering dogwood (*Cornus florida*) bloomed across the eastern United States long before European settlers arrived. For Indigenous communities, the dogwood's blooming marked the perfect time to begin planting crops. The dogwood has since become a beloved addition to gardens across North America.

One tale of its name's origin tracks back to the Celtic word *dag*, meaning sharpened pieces of wood. Valued for its durability, the timber was once crafted into tool handles, golf club heads, and other implements where hardiness was essential for use.

Botany

Dogwoods are bountiful across North America, with an estimated seventeen species thriving along the East Coast from Maine to Florida; westward to southern Michigan, Illinois, Missouri, and eastern Texas; and across the West Coast states and Idaho.

Among them, the flowering dogwood (*Cornus florida*) is perhaps the most notable, a graceful understory tree that flourishes in the shade of taller trees. It unfurls a stunning display of large snowy bracts surrounding the tiny central flowers. Flowering dogwood, a deciduous beauty, graces woodlands with its stunning blooms each spring. Native to southeastern Canada, eastern United States, and parts of Europe and Mexico, it flourishes all over the world and thrives at forest edges.

Come summertime, you can expect a flowering dogwood to transition from delicate spring blooms to a lush canopy of green leaves. In autumn, vibrant red berries appear, a final flourish of crimson before winter strips the branches bare.

The dogwood prefers soft, filtered light and slightly acidic, well-drained, fertile soil. In such a favored setting it typically reaches 10 to 20 ft [3 to 6 m]. In full sun it can reach 40 ft [12 m].

Of the estimated seventeen species, only the kousa dogwood (*Cornus kousa*) bears delicious edible fruit. The flowering dogwood bears small, bitter fruit not suitable for human consumption, but it is a treasured treat for birds, which in turn help disperse the seeds.

Pollinators

The dogwood is a vital refuge for pollinators and other wildlife, from bees to butterflies to birds and even deer. Bluebirds especially are attracted to its berries.

Dogwoods have also played a critical role in conservation efforts. In the mid-1900s, the bluebird's delicate song was fading as their populations dwindled due to habitat loss and industrial development. But since the 1980s the bluebird has made a comeback, thanks in part to the efforts of conservationists and nature lovers who cared enough to act. Following the installation of bluebird nest boxes in dogwoods across the country, as well as vigilant monitoring and protection, this charming bird has begun to reclaim its rightful place. The dogwood, with its berries beloved by bluebirds, has been a steady supporting player in the story of their resurgence.

FLOWER SHOP

Antigua

New Orleans, Louisiana

In a converted auto mechanic workshop, the New Orleans-based floral studio Antigua offers unexpected beauty. Echoes of its industrial past converge with greenery and flowers in this sanctuary where creativity blossoms.

Day to day, the studio fills with the intoxicating fragrances of fresh blooms, conjuring an otherworldly atmosphere. Here, owner Sara Perez Ekanger makes magic through the art of floral design, crafting distinctive living creations.

WHY FLOWERS?

Flowers have a way of speaking to the soul. They evoke emotions, stir memories, and remind us of life's fleeting yet profound beauty. In a world where it's all too easy to focus on the negative, flowers bring joy and light into our lives. They create connections—between people, moments, and emotions. Working with flowers humbles me. They are a live product, ever-changing and teaching me countless lessons about patience, impermanence, and gratitude. I love working with flowers because they bring joy to others and, in turn, fill my heart with purpose. This work is more than arranging blooms; it's about creating shared moments of happiness and magic.
—Sara Perez Ekanger, owner, Antigua

Part III

The Healers

There was a time when we danced in harmony with the earth. Our ancestors knew this rhythm well. They understood that the land was not to be conquered, but to be cherished and grown with. They used natural plant remedies to heal both body and soul.

In the modern world, it's easy to lose sight of all the gifts Mother Nature has bestowed on us; to move through life completely detached from the lessons imparted by every petal, leaf, stem, and root. Through celebrating the flowers in this chapter and their healing qualities, we can reconnect and embrace their power.

The healers endure as always: silent, delicate, yet undeniably powerful. Edible, fragrant, and medicinal, their benefits are endless.

In this section, we'll learn about the inner workings of these special flowers, such as how chamomile calms our senses while quietly nurturing companion plants. We'll learn how lavender, with its fragrant, feathery stems, soothes our bodies, guides us into slumber, and helps us awaken renewed. The cosmic dandelion, a reminder of our connection to the circle of life, purifies our insides while teaching us humility. The bold, radiant hibiscus offers relief from pain and inflammation. And the marigold, steadfast and bright, teaches us the power of transformation—letting go of the old to embrace something new.

These flowers embody the gentle strength of nature's medicine. They represent the resilience of life itself, patiently waiting for our return, hoping that we will remember our place in this sacred cycle. That we'll once again care for the earth as she has always cared for us. And perhaps, if we listen to nature, we'll finally, truly be healed.

Matricaria chamomilla

Chamomile

Symbols & Meanings

In the rush of daily life, it's easy to overlook ordinary simplicity as our gaze quickly shifts to grander things. If we only pause and take in our everyday surroundings, we will find the most surprising tiny treasures hiding in plain sight. Chamomile is one such wonder, with its ethereal blooms that have been prized by many for centuries.

A flowering herb symbolizing protection and an unyielding spirit in the face of life's challenges, chamomile boasts an impressive reputation. Who would have thought this modest flower could wield such quiet power?

More than just a fragrant herb, in the Middle Ages chamomile became a beloved symbol of strength and protection, believed to safeguard body and spirit.

Let chamomile remind us that even the most unassuming forces can offer profound strength and protection.

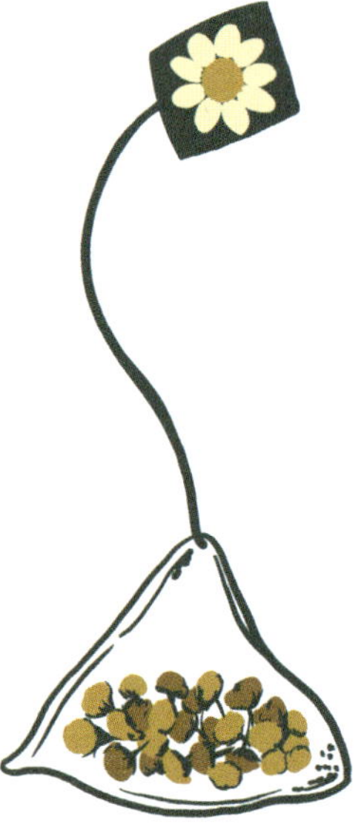

History

Chamomile's legacy stretches all the way back to ancient Egypt, where it was revered as a sacred flowering herb honoring the sun god, Ra. Chamomile became the key ingredient in the embalming oils for mummification.

Ancient Grecian and Roman cultures cherished chamomile for its remarkable healing powers and ability to restore bodily balance and calm. According to medical texts, chamomile has been used for thousands of years in traditional Chinese medicine.

Today, chamomile is enjoyed mainly as a tea, so widely drunk that an estimated one million cups are sipped around the globe every single day!

Whether for calming the mind, soothing an upset stomach, or just enjoying a leisurely, cozy moment, this flowering herb is a go-to for relaxation and comfort.

Etymology

With nods to its low, sprawling nature and crisp apple-scented blooms, chamomile's name originates from the Greek words for ground (*chamos*) and apple (*melos*).

Folklore

Chamomile's golden center hugged by blossoms has long been associated with the radiant power of the sun. In ancient Egypt, this humble bloom was not a mere herb but a sacred offering to Ra, the god who embodied the sun's life-giving power. Indeed, Ra was the sun itself.

Pollinators

Chamomile's sweet, fragrant blooms play host to a variety of beneficial insects. Though you might guess a butterfly or a bee, chamomile's chief pollinator is the beloved ladybug!

Ladybugs are model garden helpers, as they devour aphids and a variety of soft-bodied insects and larvae that can damage plants. While chamomile provides a nurturing home base, the ladybug serves as its protector, making the duo a botanical powerhouse.

Botany

There are two main species of chamomile, German chamomile (*Matricaria chamomilla*) and Roman chamomile (*Chamaemelum nobile*). While each has distinctive qualities, they share the same soothing properties, though German chamomile is most likely the one in your pantry.

Chamomile includes both annual and perennial species, thriving in climates across the globe and cultivated abundantly across Europe and in western China. While native to more tropical climates, chamomile also can flourish in colder regions.

Uses

The flowers that make up chamomile tea serve as a mild sedative, perfect for easing nervousness, relaxing muscles, relieving stress, soothing indigestion, and aiding sleep. No wonder German chamomile is included in many herbal tea blends; its naturally sweet, soft flavor contrasts with that of its more bitter cousin, Roman chamomile. Chamomile oil has long been used in various applications, from skin care to cosmetics to internal healing.

Chamomile's superior healing properties aid adjacent plant life as well as human health. It's a cherished companion plant that can help revitalize weak plants when planted nearby.

Every cup of chamomile tea carries echoes of the past, as countless generations have used it for healing and comfort. For as long as there are gardens to grow it and hearts to cherish it, chamomile's soothing touch will endure.

Recipe

STRAWBERRY CHAMOMILE COCKTAIL

This perfect summer cocktail combines the sweetness of strawberry with the floral flavor of chamomile. Fresh chamomile flowers add the finishing touch. A relaxing evening or a summer gathering is just a sip away!

MAKES 1 SERVING

1 oz [30 g] fresh strawberries

3 oz [90 ml] strongly brewed chamomile tea, chilled

1½ oz [45 ml] vodka

1 oz [30 ml] freshly squeezed lemon juice

½ oz honey syrup (equal parts honey and hot water combined)

Ice

Fresh chamomile for garnish

1. **Muddle the strawberries in the bottom of a cocktail shaker.**
2. **Add the chamomile tea, vodka, lemon juice, and honey syrup.**
3. **Fill the shaker with ice and give it a good shake to allow the ingredients to combine and chill.**
4. **Strain the mixture into a coupe glass and float a few fresh chamomile flowers on top.**

Lavandula

Lavender

Symbols & Meanings

There is a dreamy magic to lavender, an ancient knowledge threaded through each purple bloom. And though it is one of the better-known herbal flowers, that familiar presence belies its power.

Lavender, with its velvety flower heads and soothing fragrance, has long been revered for its calming qualities. For thousands of years, it's been treasured for its curative powers and practical benefits.

Lavender invokes healing, love, and renewal. By using lavender to invite peace and calm, we can find the strength to release what no longer serves us, to begin anew and flourish.

History

For over 2,500 years, lavender has been beloved for its fragrance, healing powers, and compelling allure. In ancient Egypt, its aromatic oils were treasured for both their intoxicating scent and their sacred purpose in mummification. History buffs have long speculated whether Cleopatra used lavender as her secret weapon to captivate two of the greatest Roman leaders, Julius Caesar and Mark Antony. With her wrists anointed with lavender oil, the mesmerizing scent could work its magic, with every graceful movement casting an even deeper spell.

The Greeks and Romans valued lavender's therapeutic properties. They infused it in their bathwater for its powerful restorative and relaxing qualities. Dioscorides, an esteemed Greek naturalist living in the first century, first documented lavender, ensuring its place in the annals of herbal lore.

Lavender has generated conflicting beliefs. By the Middle Ages in England, lavender had become associated with matters of the heart and was considered an aphrodisiac. Conversely, some believed it would ensure chastity if dabbed on the heads of those dearest to you. More practically, lavender was strewn across castle chambers to freshen the air and safeguard against disease.

During the Elizabethan era, when regular bathing was a rare luxury, lavender's heady aroma perfumed clothing and bed linens, transforming it into a mark of refinement. Despite its Mediterranean origins, the plant readily adapted to the English climate, earning it the name "English lavender." When the Pilgrims journeyed to the New World in the 1600s, they brought lavender along, introducing it to North America.

Centuries later, lavender's legacy took on a more symbolic meaning during the Lavender Scare—a lesser-known counterpart to the Red Scare—when suspected LGBTQ+ people were purged from US government positions nationwide based on unfounded accusations of communist ties. Activists chose the color lavender as an emblem of resistance within the community.

By 1969, as the fight for LGBTQ+ rights gained momentum, the president of the National Organization for Women infamously labeled lesbian members a "Lavender Menace," fearing they would hinder the feminist movement. Throughout the 1970s and '80s, after the Stonewall riots ignited the push for gay rights and liberation, many continued to wear lavender as a quiet yet powerful tribute to the cause—a symbol of resilience and unwavering defiance.

Today, lavender's legacy is as multifaceted as its scent: an herb of healing, a flower of enchantment, and a color of resistance and change. Whether woven into folklore, cherished for its medicinal properties, or carried as a banner of Pride, lavender continues to captivate, soothe, and inspire across generations.

Etymology

The Latin root of lavender, *lavare*, means "to wash"—a nod to its use in bathwater by the ancient Greeks and Romans.

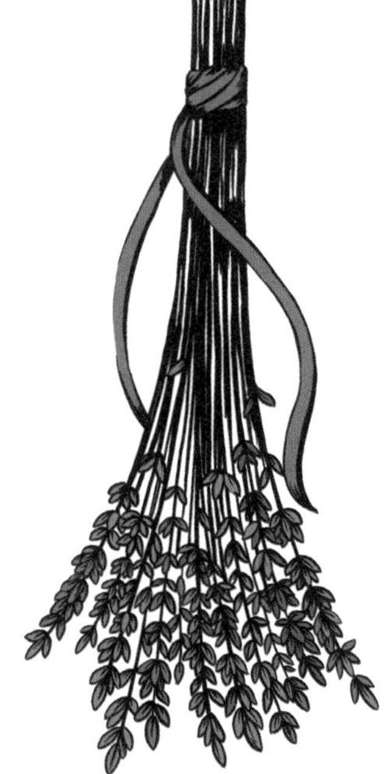

Folklore

Throughout the ages, lavender has been revered as an enchanting herb of protection, believed to ward off evil and dispel dark spirits. To guard the home, sprigs of lavender were often hung above doorways.

Beyond its protective uses, on Midsummer Eve the sweet scent was said to summon the Fae, or fairies, connecting the earthly and magical. Opening the gateway between mortal and mythical worlds, lavender became the ideal component for spells. From love spells to calming tinctures, using lavender was believed to invite love, harmony, and positive energy into one's life.

Festival

Nestled at the foot of France's majestic Mont Ventoux, the picturesque Provençal region of Sault enjoys a climate perfectly suited for cultivating lavender. Devoted growers tend the fields with reverence and artistry. Beyond its deep agricultural roots, lavender has blossomed into a cherished attraction, drawing international visitors.

Every August, the charming French town of Sault celebrates one of the world's largest lavender festivals. It opens with a traditional Provençal parade. As the festival unfolds, local artisans showcase their wares in stalls overflowing with delicate lavender bouquets, handcrafted soaps, cheeses, and an array of regional treasures. Visitors can also wander through a book fair, lose themselves in a painters' exhibition, and savor delightful refreshments.

FÊTE DE LA LAVANDE

PROVENCE

23 JULY - 15 AUGUST

Botany

The *Lavandula* genus, a mint family member, has an estimated forty-five recognized species, each offering its own distinctive charm. This resilient evergreen shrub tolerates a variety of soil types but flourishes with full sun and excellent drainage.

Invite the magic of lavender into your own home or garden and savor its many delights (see Uses). Plant the dwarf English lavender 'Thumbelina Leigh' (*Lavandula angustifolia*) where its deep purple blooms can take center stage. For those who love the art of flower drying, 'Royal Velvet' (*L. angustifolia*) is a great choice.

Uses

While lavender is renowned for its soothing topical applications, its medicinal benefits are equally remarkable! Scientific studies have shown that inhaling lavender essential oils can reduce anxiety and also act as a sleep aid. The potent natural compounds linalool and linalyl acetate are the key constituents responsible for lavender's tranquilizing effects. Keep a bottle of lavender essential oil on your bedside table for some middle-of-the-night aromatherapy when sleep eludes you.

Culinary species of lavender make a delightful addition to any herb garden. For savory dishes, 'Provence' (*L. x intermedia*) pairs beautifully with meats, infusing a subtle, herbaceous flavor. For a sweet treat, 'Melissa' (*L. angustifolia*) lends a gentle floral note that enhances the natural sweetness of baked goods.

Recipe

LAVENDER SUGAR SCRUB

This DIY scrub is a simple project with big benefits. The combination of the sugars and beneficial oils will exfoliate and moisturize rough, dry hands and feet, leaving your body pampered and your senses relaxed. The fragrance will make you feel like you are frolicking in the French countryside surrounded by rows and rows of lavender.

MAKES 2 CUPS [480 G]

½ cup [120 ml] coconut oil

1 cup [200 g] raw sugar

1 cup [200 g] packed brown sugar

1 Tbsp dried lavender buds

10 to 15 drops lavender essential oil

A few drops of vitamin E oil (optional, for extra skin-loving benefits)

Lavender buds (optional, for decoration)

Two 8 oz [240 ml] jars with lids, sterilized

1 **Melt the coconut oil slightly if it's solid, just until soft.**

2 **In a bowl, combine the sugars and dried lavender by hand.**

3 **Add the coconut oil and stir until completely combined.**

4 **Add the lavender essential oil and vitamin E oil and blend well.**

5 **Spoon into the jars. You can decorate the lid by gluing or tying a few additional lavender buds onto it. Voilà! A beauty product with minimal and pure ingredients.**

6 **To use: Indulge in a gentle massage on damp skin, using circular motions, then rinse to uncover velvety soft skin. Smooth, supple, delicately perfumed skin—without even leaving home!**

Taraxacum officinale

Dandelion

Symbols & Meanings

The dandelion is often overlooked, its humble beauty and extraordinary qualities rarely given the recognition they deserve. Often dismissed as a mere weed, it actually embodies resilience and nourishment.

The dandelion stands as a testament to the power of perseverance and triumph over life's challenges. Whether on rocky hillsides or in sidewalk cracks, the dandelion flourishes—a radiant reminder that no matter the darkness surrounding us, sunlight will always return, offering brighter days ahead.

And let's not forget the dandelion's most enchanting gift: the ability to make wishes come true. With a soft puff of air, its seeds scatter to the wind, carrying our hopes and dreams.

Folklore

Sometimes dubbed “fairy clocks,” dandelions’ golden faces open with the dawn and gently close at dusk. This rhythmic pattern was once thought to help keep the fairies on time—hence the nickname. When you make a wish and launch a cloud of dandelion fluff, legend has it that fairies carry your whispered desires to the winds, guiding them to fulfillment.

The dandelion’s magic even stretches beyond earthly bounds. An old wives’ tale proclaims that it represents the celestial cycle: Its golden bloom mirrors the sun, its billowy seed head glows like the moon, and when its seeds scatter, they resemble stars dancing across the night sky. With this connection to the cosmos, dandelions remind us that even the smallest wonders are part of something vast and timeless.

Etymology

The name dandelion is derived from the French *dent de lion* (“lion’s tooth”), a nod to the fanglike jagged edges of its leaves.

Botany

This dandy lion is more like a pussycat than a lion, brimming with vitality, offering vitamins and minerals that nourish the earth and all who dwell upon it.

The dandelion is an incredibly resilient plant. This humble flower thrives through a successful strategy: emerge and bloom briefly, disperse seeds widely, send roots deeply. In exchange for fertile, well-draining soil, the dandelion gives much in return. Its deep roots loosen compacted earth, inviting the presence of life-giving earthworms, which enrich the soil with their vital work.

The dandelion is ubiquitous through its remarkable seed dispersal. A single plant can produce nearly twenty thousand seeds, which float on the breeze to take root and flourish in new ground. Seeds germinate quickly, ensuring the next generation. The plant's roots, capable of reaching an impressive fifteen feet deep, can regenerate when divided. So the next time you make a wish on a dandelion puff, realize you're contributing to the species' propagation!

As if we needed more reasons to love this tough powerhouse of a plant, the dandelion is also a cherished resource for pollinators. By welcoming dandelions into our spaces, we enable an environment that nourishes wildlife and strengthens nature's delicate balance. These "weeds" are far superior to many of our carefully curated garden plants, offering a bounty of benefits to both the earth and the creatures who call it home.

History

Though the dandelion is found across the globe, its healing properties have long been revered by herbalists practicing traditional Chinese medicine. Arabic peoples later embraced the dandelion's medicinal virtues, and before long the flower found its way into kitchens and apothecaries across Europe, becoming a trusted staple.

While some dandelions are native to North America, early European settlers brought with them the two types most well-known on the continent today: the red-seeded dandelion (*Taraxacum erythrospermum*) and the common dandelion (*Taraxacum officinale*). These settlers introduced the flowers for both their beneficial uses and their ability to thrive even in compacted, poor-quality soil. Over time, these resilient plants have become naturalized, with dandelions now flourishing throughout the United States.

Uses

The dandelion is a beloved remedy for ailments both small and large, though many of its uses in medicine are not studied or backed up by modern science. Its roots and leaves are believed to have anti-inflammatory properties: Dandelion tea supports kidney function (as a gentle diuretic) and liver and gallbladder health, eases muscle tension, and alleviates headaches. The milky sap sourced from the stem can be used as an ointment and is believed to be a remedy for warts, corns, moles, pimples, calluses, sores, bee stings, and blisters.

Dandelion greens are also a potent nutritional ally in maintaining health and vitality. Rich in a spectrum of essential vitamins—including A, C, E, and K—they also boast an impressive mineral profile—potassium, iron, and magnesium—and fiber content.

Craft

WISHES IN A JAR

This simple, beautiful project captures the magic of dandelion wishes and preserves them in a whimsical way. It's not only a wonderful method of preserving dandelions' fleeting allure but also a heartfelt reminder that wishes, just like seeds, have the power to grow.

Dandelion seed heads (the puffs!)

Scissors or clippers

Small piece of paper

Pen

A clean, dry Mason jar

6 in [15 cm] square piece of fabric

Rubber band

Twine

1. Gather dandelion seed heads (the fluffy white puffballs). Cut the stems from the plant's base carefully, so the fluffy parachutes do not blow away as you gather them.

2. `Write a wish or wishes (optional):` On the small piece of paper, write down a wish, hope, or dream that you would like to come true. (You can either fold the paper and tuck it into the jar or punch a hole to string it onto the twine in step 5.)

3. Cut off the stems and carefully place the dandelion seed heads in the jar, letting the delicate fluff fall gently inside. Leave the seeds intact, allowing them to float freely within the jar.

4. With the seeds safely inside, center the fabric piece over the top of the jar and secure it with the rubber band.

5. Tie a length of twine around the fabric to cover the band.

6. `Display your dandelion wishes:` Place the jar somewhere special in your home—a shelf, windowsill, or table—as a reminder of the magic of your wishes and the joy you'll feel as they come true.

Hibiscus

Symbols & Meanings

In North America, the hibiscus has long been associated with the idealized woman and the epitome of feminine beauty. During the Victorian era, offering a hibiscus flower was a tender gesture, an unspoken acknowledgment of the recipient's mesmerizing loveliness.

In China, the hibiscus carries a bittersweet message. Representing the fleeting nature of fame and glory, the hibiscus acts as a cautionary tale, reminding us that beauty and recognition are as temporary as flowers.

For the women of the Pacific Islands, the hibiscus takes on a deeply personal meaning, often worn to signal their relationship status. If she was open to romance, she would tuck it over her right ear; if already in a relationship, she'd it place it over her left ear.

Home

The yellow hibiscus (*Hibiscus brackenridgei*), also known as maʻo hau hele, is the reigning official state flower of Hawaiʻi. Although the hibiscus was designated the official flower of the Territory of Hawaiʻi in 1923, the specific species was left undetermined, so many people believed the native red hibiscus represented the state. Finally, in 1988, the Hawaiʻian legislature officially designated the maʻo hau hele as the true state flower.

The maʻo hau hele thrives solely on the main Hawaiʻian Islands. Found in the dry forests and shrublands, the maʻo hau hele graces elevations between 400 and 2,600 feet, quietly adding its beauty to the islands' rich tapestry and heritage.

Revered in Hawaiʻian culture, the bloom has also been used to craft a distinctive blue-gray dye for the traditional kapa cloth. Rare and deeply connected to the land, this flower is considered endangered, and conservationists are taking significant steps to preserve this treasured part of Hawaiʻi's unique landscape.

History

When you hear "hibiscus," you probably envision *Hibiscus* x *rosa-sinensis*, the Chinese or tropical hibiscus. The *Hibiscus* genus boasts hundreds of species thriving in tropical and warm climates, yet their wild counterparts bear little resemblance to the flower we associate with the name.

Centuries of careful hybridization across diverse regions have shaped the hibiscus into the flower we know today. In recent years, innovative breeding techniques have further refined the hibiscus, transforming these once spindly landscape plants into robust and cherished specimens for home gardens.

Unlike many botanical species, *Hibiscus* x *rosa-sinensis* has not been discovered in the natural world, yet its origins almost certainly trace back to Asia. Though *sinensis* translates as "Chinese," India is widely considered the true birthplace of the original hibiscus; early travelers from the Indian subcontinent carried this beautiful bloom to the South Pacific islands, where its captivating charm was then shared with the rest of the world.

Regardless of its origin, China nurtured and cultivated the hibiscus. European explorers marveled at the diverse varieties flourishing around ancient temples, and soon carried the hibiscus to Europe. The hibiscus is also entwined in the religious and cultural traditions of Malaysia, Tahiti, Fiji, and Hawai'i.

Botany

Hibiscus flowers include both tropical and hardy species. Each type offers its own distinctive charm, and the right choice will make an enchanting addition to any garden!

The hardy hibiscus (*Hibiscus moscheutos*) stands tall, with dramatic, oversized blooms, gracefully enduring even the coldest winters. In contrast, the tropical hibiscus (*Hibiscus* x *rosa-sinensis*) cannot withstand winter chill but opens its vibrant, exotic petals above glossy leaves in warmer climes. The rose of Sharon (*Hibiscus syriacus*) is a beloved garden staple, though considered invasive in some regions. Its blooms may be smaller than showier species', yet they hold their own in beauty, rivaling the grandeur of the other varieties.

The art of hibiscus breeding began in Hawai'i, then extended to the mainland, resulting in the 1950s formation of the American Hibiscus Society. Dedicated to education and preservation, this group sought to establish a system to track, record, and register the ever-growing array of *Hibiscus* x *rosa-sinensis* cultivars. Today, this important work continues, ensuring the legacy of this beloved bloom.

Pollinators

Butterflies and hummingbirds are drawn to the hibiscus' striking colors and promise of sweet nectar. Hummingbirds have a special affinity for the popular crimson shade.

Uses

The magic of hibiscus extends beyond its beauty and allure in nature. Hibiscus tea, treasured for its vibrant red color and soothing properties, is made from *Hibiscus sabdariffa*, also known as roselle or Florida cranberry. Despite the name, this species is native to West Africa, from where it has spread across Central America, the Caribbean, and Florida. Long used to ease headaches, soothe aches and pains, and calm coughs and inflammation, hibiscus offers comfort, one sip at a time.

Folklore

Kali is one of Hinduism's most formidable goddesses, embodying both destruction and rebirth: life's eternal cycle. Her fierce appearance, with a garland of severed heads and swords in her many hands, reflects her fearless and relentless battle against darkness and malevolence.

Yet beneath this daunting exterior lies deep compassion. Kali is a force not only of annihilation but also of transformation and renewal, clearing the path for growth and liberation. She represents the divine feminine energy in its most powerful form, giving us the confidence to confront our fears and overcome the obstacles that threaten our peace and progress.

In sacred rituals dedicated to Kali, Hindu practitioners offer her red hibiscus flowers—their vivid hue mirroring her fiery energy—in hopes of enlightenment and fulfillment of their deepest desires.

Flowers hold a sacred place in Hinduism and are often woven into local worship. In the tradition of puja ("flower prayers"), flowers are lovingly presented to the deities as tokens of devotion and respect. Worshippers believe that through these acts, divine favor is bestowed, bringing abundance, protection, and spiritual awakening.

Tagetes

Marigold

Symbols & Meanings

The marigold, with its radiant yellow, orange, and red hues, is strongly associated with the sun's energy and its remarkable power of resurrection. Marigolds have been used as offerings to deities across various cultures, honoring their vibrant life force.

In Buddhism, marigolds are revered as offerings to Buddha; in Hinduism they are linked to the sun, bestowing blessings and joy during marriage ceremonies. In the Middle Ages these golden blooms were woven into love charms.

In the Victorian era, however, the marigold took on a more sorrowful meaning, representing despair, grief, and the pain of lost love. But while marigolds have long symbolized death, the afterlife, and transformation, their presence is far from ominous. They carry the wisdom of change, adaptability, and the unknown pathways of fate. Marigolds honor death not in sorrow but in remembrance and resurrection of spirit.

These golden flowers have been cherished for centuries as symbols of the departed, their bright petals lighting the way for the souls of the dead during celebrations like Día de los Muertos. Marigolds remind us that in death there is not only an end but the beginning of something new, a transformation into something beautiful, and the wisdom to embrace the future with grace.

History

Marigolds, native to South America, Central America, and Mexico, were cultivated on the chinampas—ingenious floating agricultural islands that once surrounded the city of Tenochtitlán, where the Aztecs incorporated them into medicine, rituals, and daily life.

As Spanish colonization began, marigolds were frequently traded items. They soon graced the gardens of Europe and later the United States.

In the sixteenth century, Portuguese merchants brought the marigold to India. There they quickly became the flower of choice for temple offerings, weddings, and the luminous festival of Diwali, the Hindu festival that honors the triumph of light over darkness. Garlands of marigolds adorned homes and inspired the creation of rangolis, intricate works of art crafted on the floor with brightly colored sand, rice, flowers, and more to mark the holiday's sacred and celebratory nature.

Illuminating life's most sacred moments, from the floating gardens of the Aztecs to the grand festival of Diwali, this bloom is forever immortalized in human history.

Uses

Marigolds have been trusted for centuries in folk medicine—they are little powerhouses of natural healing!

Marigolds contain cancer-preventive, antioxidant, and anti-inflammatory powers to soothe the body inside and out. They also lend a helping hand to the digestive system, keeping things running smoothly.

With just as much vitamin C as tomatoes, marigolds help boost the immune system and keep skin glowing. They're also loaded with potassium and contain more iron than spinach. Drink it in a tea or add as a garnish to make any heart happy.

Folklore

The marigold holds a sacred place in the traditions of Día de los Muertos, a celebration originating in Mexico and enduring as one of the country's most cherished rituals. For years, marigolds have unfurled their golden petals across Mexico, their brilliance woven into the heart of this tradition. Their vivid colors cascade over altars and adorn crucifixes, an inspired tribute to the eternal bond between the living and the dead.

As November 1 approaches, families gather in remembrance, adorning gravesites and home altars (*ofrendas*) with these radiant blooms. It is believed that marigolds' alluring scent and bright petals light a symbolic pathway, welcoming departed loved ones to return home to be celebrated.

The marigold's petals are also believed to hold cleansing powers and thus are often used to create a cross in front of the ofrenda. This cross serves as a sacred space where the souls may step and be cleansed of their earthly sins, gaining absolution and peace.

Botany

These radiant annuals are a treasure for any garden. Marigolds offer many benefits that provide protection from pests and create a delightful home for pollinators! They flourish in light, well-drained soil and full sun, preferring dryness to excess moisture and damp.

Belonging to the aster family and the *Tagetes* genus, marigolds fill gardens with a treasure trove of warm-toned hues throughout the summer and fall. The three main varieties—French, African (or American), and signet—each offer distinctive charm and multi-colored hues.

Marigolds exhibit more than just beauty. Their scent wards off unwelcome visitors such as deer and rabbits while inviting a host of beneficial insects that naturally combat pesky garden pests. Certain marigold varieties can resist some nematodes—microscopic worms that ravage plant roots. Some marigold roots release a chemical that makes an uninhabitable environment for nematode eggs, a powerful form of pest control. With their beauty and resilience working in tandem, marigolds enrich our gardens in ways both seen and unseen.

Pollinators

While the strong scent of marigolds is effective at deterring many unwanted insects and protecting garden plants, these vibrant flowers also play an important role in supporting beneficial pollinators.

Marigolds offer a delectable feast for caterpillars, whose transformation into butterflies and moths brings enchantment to the garden; they are a habitat for certain Lepidoptera caterpillars and offer a rich nectar source for butterflies, moths, and honey bees. The bright, electric hues of marigold blooms are especially attractive to these pollinators, drawing them in from afar. Including marigolds in your garden not only adds color and charm but also helps sustain essential pollinator populations.

Craft

HOW TO DYE WITH MARIGOLDS

Marigold flowers are not only beautiful but also wonderful for creating rich golden-yellow or olive-toned dyes. The process is simple and natural, making it a perfect craft for those looking to bring a touch of nature into their artwork or fabrics. For best results when dyeing fiber with marigolds, aim to use equal weights of marigold petals and dry material (fabric or yarn). A good kitchen scale is a must for calculating the weight of this and other ingredients.

Note: As with all dyeing crafts, wear rubber gloves and an apron, and protect surfaces with sheets of plastic or tarps. Designate a large pot that will not be used for cooking, and ensure good ventilation.

Fabric or yarn (natural fibers like cotton, silk, or wool work best)

Pot large enough to fully submerge the materials

Alum, equal to 15 percent of material weight

Cream of tartar, equal to 5 to 6 percent of material weight

Fresh or dried marigold flowers, equal in weight to the fabric used

Strainer

1 **To prepare the material:** Wash fabric or yarn thoroughly to remove any dirt, oils, or chemicals that could prevent the dye from adhering properly.

2 Fill a large pot with enough warm water to submerge the material. Dissolve the alum and cream of tartar and add the material. Bring the water to a simmer and let the material rest in the pot for 1 hour at a consistent simmer.

3 Remove the material and wring it out, reserving the pot of alum water.

Note: Whether using fresh or dried marigold flowers, you can use whole tops or pluck off and reserve the petals; either way will release the dye.

4 **To create the dye bath:** Add the marigold petals to the alum water. Bring the mixture to a boil, then lower the heat and let simmer until it reaches your desired color, about 30 minutes to 1 hour. The petals will release their vibrant golden color into the water. Stir occasionally to make sure the dye is evenly distributed.

5 Once the dye has reached your desired color strength, strain out the marigold petals using a fine mesh strainer. You now have a clean dye bath ready for the fiber.

6 **To dye the material:** Carefully submerge the prepared material in the dye bath, making sure it's fully immersed. Keep it in the dye for 30 minutes to 1 hour, depending on how deep a shade you want. Stir the fiber occasionally, releasing any folds to ensure full distribution of the dye. The longer you leave the material in the dye, the more intense the color will be. For a lighter shade, remove the material after 20 to 30 minutes.

7 Once you're happy with the color, remove the material from the dye and rinse it in cool water until the water runs clear. This both removes excess dye and helps set the color. Hang the material to dry in a shaded, well-ventilated area. Do not dry it in direct sunlight, as this could fade the color.

8 After the dyed material has dried, you'll have a beautiful, natural marigold-dyed piece. Whether it's a scarf, pillowcase, or skein of yarn, the rich golden or olive hue from the marigolds will add a lovely touch of nature to your home or wardrobe.

FLOWER SHOP

ISLAND-BOY Shop

Honolulu, Hawai'i

At ISLAND-BOY shop, tucked away in Honolulu's Kaimuki neighborhood, you're instantly drawn to the eclectic curation of items. The centerpiece is an ever-changing floral arrangement inspired by Hawai'i's unique environment, using locally sourced stems such as protea, anthurium, and other tropical flowers.

Given Hawai'i's abundance of flora and unique gift-giving culture, you'll find that a wide selection of leis—garlands of flowers, leaves, vines, shells, seeds, nuts, or feathers—are central to Hawai'ian life. Leis are beautiful, tangible representations of the time and warmth used to make them, imbued with the best of intentions from maker to recipient. It's customary to gift loved ones with leis to honor the interconnectedness of life and celebrate significant milestones.

Andrew Mau, the owner of ISLAND-BOY shop, makes leis from flowers grown and picked by family and neighbors. These reference tradition but offer innovative designs and unexpected combinations. Andrew also hosts lei-making and floral-arranging workshops in collaboration with Pālehua Conservation Initiative (PCI), a nonprofit that protects native resources by educating the public on Hawai'ian ecology.

WHY FLOWERS?

When I first started to work in furniture and interiors, I learned that plants and flowers are utilized to add life, color, and personality to a space. A room without them can feel so lonely. We try to always have flowers around the shop, so shoppers feel welcomed in the space. In a way, they sort of keep you company.

—Andrew Mau, owner, ISLAND-BOY shop

Part IV

The Wild Ones

Say hello to nature's ultimate survivor: the wildflower, thriving in fields, forests, and even the cracks of a hot city sidewalk. Each variety of wildflower is unique and hardy, surviving with few resources, self-seeding and self-sustaining, spreading without the helping hands cultivated flowers depend on.

Wildflowers evoke happiness, loyalty, optimism, honesty, and longevity. They are the world's oldest teachers, speaking lessons of resilience and hope. All we must do is listen.

Stop. Breathe in. Look around and listen. The world beckons us to hustle and bustle, but if you pay attention, nature is quietly imploring us to slow down.

Wildflowers hold an interesting space in our world—a contradictory blend of ordinary and otherworldly. Have you ever listened to the busy hum of bees working over a purple coneflower plant in the summer? Or perhaps you remember violets being the first flower to emerge from the icy ground in your grandmother's yard after a grueling winter. Maybe you blew on the silky tendrils from a milkweed pod, making wishes. You might've plucked the petals from a field-picked daisy, divining whether your crush loves you—an adolescent's rite of passage. Nature continually provides us with the magic we long for, and wildflowers play a key role.

Tap into this childlike wonder of the wildflower. Untamed. Free. Resilient. Leaving things a little better than we find them. Harness the power of being common.

Asclepias syriaca

Common Milkweed

Symbols & Meanings

Milkweed growing in the garden can symbolize remembrance, protection, dignity, and freedom. Its genus, *Asclepias*, is named after the Greek god of healing. Milkweed has been used as a painkiller and a pulmonary aid, just to name a few.

Botany

Common milkweed grows wild in most of the eastern United States, and along with other native plants, it provides a much-needed support system for the animals and insects that pollinate our food and provide pest control.

After flowering, the milkweed plant's next job to ensure survival is to disperse its seeds. The plant creates a white fluffy substance called floss that encompasses the seeds; think the dandelion's cooler older sister. The floss, along with the seeds, gets carried away by the wind until it lands to germinate, take root, and grow into an entirely new plant.

Pollinators

Monarch butterflies (*Danaus plexippus*) have one of the longest migrations of any butterfly on the planet. Every autumn, millions of monarch butterflies migrate two thousand miles from their breeding grounds in North America and Canada to spend the winter in the mountain forests of Mexico. That's a lot of mileage for those beautiful wings.

Monarchs depend on the nourishment of all milkweed flowers—not just common milkweed—to fuel their journey, and luckily, this beautiful wildflower grows in many different colors throughout fields in the United States. Milkweed not only feeds them but also hosts the monarch's entire life cycle, aside from the chrysalis stage, offering a home for eggs to hatch into larvae. Hungry caterpillars eat the leaves, building strength to shed their outer layer, so nature's most magnificent transformation—metamorphosis—can take place. Finally, a beautiful monarch butterfly emerges.

Monarchs and their migration are currently threatened by habitat destruction, global warming, and certain farming practices in the United States. Millions are disappearing. They can be helped by planting milkweed in your backyard and by avoiding pesticides. Three milkweed species have particularly wide ranges and are good choices in most regions: common milkweed (*Asclepias syriaca*), swamp milkweed (*A. incarnata*), and butterflyweed (*A. tuberosa*).

Folklore

Legend has it that if you catch milkweed floss, make a wish, and let it go, your wish will be granted.

Milkweed and fairies are inseparable, and not just because they *look* like they belong in the same enchanted storybook—they actually do! These two have been linked in tales for ages. In Victorian-era stories, we're told of butterflies being fairies in disguise. Milkweeds are also known as a protective shelter plant for fairies, according to Fae culture, which has its magical roots in various European traditions, drawing inspiration from Celtic myths, Slavic folklore, Germanic legends, and the tales of the French countryside.

In these traditions, it's believed that the Fae, or fairies, drift through dreams, much as milkweed seeds gracefully drift on a breeze, guiding, guarding, and sometimes gently meddling.

Stuffing a dream pillow with milkweed floss will help you dream of the Fae (see Making a Dream Pillow, page 140).

Crafts

HARVESTING AND REPLANTING

Harvesting milkweed seeds is not only a fun project to turn to when you have some free time but can also make you an active participant in the species' survival.

Begin by monitoring the pods that the plant produces. Once they begin to split, secure them with a rubber band to prevent the pods from opening entirely. Next, take the pods inside and collect the seeds. The harvested seeds can be kept in the refrigerator until it's time to plant them in the spring.

A bonus? When they bloom, maybe some monarchs will flock to your yard.

MAKING A DREAM PILLOW

Invite fairies into your dreams with a dream pillow stuffed with milkweed floss—a bedtime magic filled with herbs and good intentions. Fairies help summon sweet, meaningful dreams while standing guard to keep nightmares at bay.

Fairies fluttering into our dreams are meant to invite magic and joy into our lives, reminding us to not take life or ourselves too seriously. Life is meant to be enjoyed, savored, and cherished!

Fairies are intertwined with nature—they will communicate through flowers or a bird's sweet song to encourage us to tune in to our intuition. It is always wise to pay attention to the natural elements in a dream state—it could be the fairies trying to tell you something! Dare to follow the dream fairy to reach a place of clarity or a moment of euphoria.

Creating a dream pillow by stuffing it with milkweed floss will invite this magic into your dreams. Allow it to awaken your senses, and enjoy the journey.

A handful of milkweed floss, seeds removed

A sprinkle of lavender

Two 8 in [20 cm] square pieces of thin cotton fabric

Needle and thread

1. **Mix the milkweed floss and lavender together to make the stuffing.**

2. **Stack the fabric squares inside out. Sew three sides together, using a 3⁄8 in [1 cm] seam allowance.**

3. **Partially sew the fourth sides together, leaving a 2 in [5 cm] opening. Using that opening, turn the pillowcase right side out. Use a pencil or chopstick to push out the corners.**

4. **Loosely fill the pillow with the milkweed floss and lavender mixture without overfilling. Tuck the raw edges of the opening inside the pillow and sew that opening closed. Trim off the extra thread.**

5. **Tuck your dream pillow under your bed pillow and embark on a night full of magic.**

Festival

Every year the town of Bethel, Maine, hosts a Monarch Festival, a citywide celebration organized by the Mahoosuc Land Trust. Residents and visitors gather, planting milkweed to attract monarchs on their migratory journey to winter in Mexico. Bethel's lush, serene setting of rolling hills provides the perfect stopping point for these winged wonders, a place to rest and recharge. The event is a testament to the community's commitment to conservation and education, and a celebration of the natural world.

CELEBRATE NATURE'S MAGESTIC MIGRANTS

PLANT MORE MILKWEED • SAVE MONARCHS

Echinacea purpurea

Purple Coneflower

Symbols & Meanings

The purple coneflower's drought resistance and adaptability have made this gardeners' favorite a lasting emblem of endurance. Its resilience in the toughest situations has made it a symbol of strength and healing.

Botany

The purple coneflower, or echinacea, is one of the United States's most iconic and enduring native species. Purple coneflower grows wild in prairies, meadows, and wooded areas in much of the central and southeastern United States. This colorful bloom has a rich and storied history.

Etymology

The purple coneflower's botanical name, "echinacea," comes from the Greek *echinos* ("hedgehog") for the flower's round, spiky central cone.

History

The Indigenous American Ute people referred to the purple coneflower as "elk root" because of the animal's tendency to dig up their roots. Not only elk, however—various livestock animals have been spotted gathering up purple coneflower roots in their grazing paths.

Home

The Tennessee coneflower (*Echinacea tennesseensis*) has earned the title of one of Tennessee's state wildflowers. It is one of the few plants exclusive to the dry, limestone-filled terrain of Tennessee. At one time it was thought to be extinct, only to be miraculously rediscovered in the late 1960s. Through conservation efforts like land extension and preservation, the species was revived.

Uses

Echinacea, as the dried flower is called, is well-known for its medicinal properties, including as a remedy for inflammation and pain and as an immunity builder. A tea made from its roots is commonly sold pre-made in grocery stores, a testament to the plant's healing power.

Folklore

In Native American cultures, the purple coneflower is a botanical miracle, a cure-all used to treat ailments for centuries. The healing properties of echinacea are long discussed, as well as the strength it exhibits in the garden.

Recipe

ECHINACEA SYRUP

This anti-inflammatory and immune-boosting concoction will soothe that pesky common cold and speed recovery.

1 oz fresh ginger, grated

1 oz dried echinacea root

½ cup [170 g] honey

1. To make the tea concentrate: Add the ginger and echinacea root to a pot with 1 pint of cold water. Bring to a boil and simmer on low until reduced by about half. Strain out the solids and reserve the liquid.

2. Gently heat the honey and tea concentrate together, stirring frequently, until fully combined.

3. Let cool and transfer to a Mason jar or bottle with a lid. The syrup will keep for up to 3 months, sealed snugly and stored in a cool, dry, dark place.

Pollinators

Adding purple coneflower to your garden will boost pollination and round out a vibrant ecosystem. Bees and butterflies swarm to the summer flowers. And in fall and winter the goldfinch, while not technically a pollinator but a seed disperser, and other wintering seed-eating birds enjoy this plant long after the beauty has faded. Leave the dead stalks standing through fall and winter so these bright birds can continue to feed on the ripe seeds. (The finches' winter plumage is more subdued but still accented with black-and-white wing bars.)

Kalmia latifolia

Mountain Laurel

Symbols & Meanings

Mountain laurel has earned its symbolic badge of perseverance due to its ability to thrive in so many places—across the eastern United States from Maine to Florida and west to the Mississippi.

Botany

This evergreen shrub is an excellent choice for the shady garden. Its twisted trunks appear taken straight from a fairy tale—perfectly imperfect, growing this way and that. In late spring and early summer, clusters of beautiful flowers bloom, ranging from white to pink and marked with distinctive spots or streaks. After the flowers fade, the dark green foliage endures through the colder months.

History

Mountain laurel has a long history in America. The wood is perfect for carving; durable and beautiful, it takes shape well and holds up strongly. This quality earned the mountain laurel the nickname "spoon wood" from certain Native American communities.

Mountain laurel is the state flower for not one but two states: Connecticut (recognized in 1907) and Pennsylvania (1933). Tourists flock to the mountains each spring to enjoy the mountain laurel's blooming from late May well into June.

Folklore

The mountain laurel is associated with ambition—and ambition rewarded. The expression "resting on your laurels," which refers to a satisfaction with past achievements and no desire to gain new ones, dates back to ancient Greek and Roman traditions. Victorious athletes or generals were crowned with laurel wreaths to represent their victory, success, and status. The Greeks presented a laurel wreath to poets, athletes, and war heroes as a mark of great achievement.

Greek mythology tells of Apollo (god of light, medicine, music, art, and archery) and his reluctant love interest, Daphne. As she fled from the lovestruck Apollo, Daphne was transformed into a laurel tree.

Pollinators

The intricate flowers of mountain laurel attract native bees and hummingbirds like no other shade-blooming shrub. And no wonder: The mountain laurel's syrupy flowers are special. The plant delegates the work of releasing its pollen elsewhere: Before the flowers open, the anthers, or pollen carriers, are sheltered from the elements, developing while the petals are closed. When the petals open, they increase tension on the anthers, essentially creating tiny catapults. So in a way, when the flower blooms, the anthers do, too. Then, when a bumble bee flies by searching for nectar, the anther rubs up against the bee's body, coating it in pollen, so the bee spreads it like fairy dust all along its flight path. The efficiency of placing the pollen on the pollinator is just another one of nature's genius tactics.

Festival

Every year in June, the community of Wellsboro, Pennsylvania, celebrates the seasonal bloom of the state flower during the Pennsylvania State Laurel Festival. This week-long celebration includes several events lavishing praise on the mountain laurel and encouraging general community merriment, from an arts and crafts fair to foot races, a pet parade, a traditional parade, a gospel performance, several concerts, and a pageant where the Pennsylvania State Laurel Queen for the year is chosen.

The festival was started in the 1930s by local businessman Larry Woodin to promote the nearby Pine Creek Gorge, also known as the Grand Canyon of Pennsylvania.

THE HISTORIC APPALACHIAN
Mountain
Laurel
Festival
EST 1938

Aquilegia canadensis

Wild Columbine

Symbols & Meanings

In medieval times, columbines represented foolishness because of their similarity to a court jester's slippers.

Botany

Wild columbine, also known as eastern red columbine, is an exquisite wildflower with bell-like flowers and backward-pointing spurs that store the nectar.

A member of the buttercup family (Ranunculaceae), wild columbine is evergreen. Unless the temperature exceeds 110°F [43°C] or dips below –10°F [–23°C], the plants remain green and ready to bloom in the spring.

Etymology

The genus name, *Aquilegia*, comes from the Latin word *aquila* ("eagle"), referring to the claw-like spurs that suggest an eagle's talons.

Columbine derives from the Latin *columba* ("dove"), as the back of the flower is said to resemble a circle of doves drinking around a fountain.

History

This native wildflower has thrived in both America and Europe since the 1600s. The genus *Aquilegia* has been determined to originate in Eastern Europe and Central Asia. The plant reached North America during the Pleistocene, somewhere between ten thousand to forty thousand years ago. During this time, the Bering Land Bridge connected Asia to part of North America. This land connection made it possible for plants and animals to spread between the continents.

Folklore

For all sorts of afflictions from heartache to headache and more, the wild columbine plant offered a cure for Native American communities. It was believed that crushed seeds were the perfect ingredient for a love spell and cure for a broken heart. Native American men would crush the plant's seeds and cover their hands in its powder as a charm to entice romantic partners.

Pollinators

The wild columbine's blooms attract a variety of pollinators, including hummingbirds, bees, butterflies, and hawk moths. The hollow spurs provide a unique passageway to the plant's nectar, which can be reached only by long-tongued feeders.

The ruby-throated hummingbird is especially partial to the plant, relying on its sustenance during spring migration northward. The flower's extraordinary shape is both purposeful for the pollinators and a pleasure for spectators. And the ripe seeds of the plant are eaten by finches, buntings, and other seed-loving birds.

Festival

Every year in Oregon's Awbrey Park in Eugene, there's an annual celebration dedicated entirely to wildflowers. The Friends of Awbrey Park, a neighborhood volunteer group that adopted the park and provides caretaking, station themselves throughout the park during the festivity. They can help you find plant wonders like trillium, lupine, red columbine, bleeding heart, violets, fairy bells, false Solomon's seal, Oregon myrtle, and so much more. Guests can take a self-guided tour and discover the native flowers, shrubs, and trees throughout the park. The celebration wraps up with a native plant treasure hunt: The winner takes home their very own potted native plant.

Oregon's
Wildflower
Festival
Aubrey Park

Rosa virginiana

Wild Rose

Symbols & Meanings

With its delicate petals and sweet aroma, the wild rose has become synonymous with love, beauty, adoration, and spiritual transformation. Roses are associated with Aphrodite, the Greek goddess of beauty and love.

Botany

The lovely plant known as the Virginia rose, common wild rose, or prairie rose (*Rosa virginiana*) is native to eastern North America. This vigorous perennial shrub reaches 6 ft [183 cm] tall, its stems covered with numerous hooked prickles. Small clusters of pink flowers bloom in summer, and the glossy green foliage turns a brilliant yellowish red in the fall. It thrives in full sun to partial shade in a variety of soil types with good drainage. *Rosa virginiana* provides food for many forms of wildlife, and its ability to tolerate high levels of salt makes it a good choice for coastal gardens.

Uses

The rose hips, famously rich in vitamin C, can be eaten, steeped to make tea, and used in homemade beauty products. The plant's roots have been used in baths, for treating cuts, and as a soothing eye wash.

History

Rosa virginiana has a rich history in North America dating all the way back to 1725, when early New England settlers found the native roses appealing and sent them back to their loved ones in England.

Pollinators

This common rose attracts a wide array of wildlife that depend on the plant for food and shelter. A variety of native bees nest beneath and within the plant and harvest parts of the plant to construct nests elsewhere.

These roses offer stomping grounds for caterpillars, and their dense shrubbery provides a safe habitat for birds and small mammals. The blooms' pollen is enjoyed by flies, beetles, and over one hundred species of butterflies.

Festival

The Elizabeth Park Conservancy's rose garden in West Hartford, Connecticut, is the nation's oldest municipally operated rose garden, boasting nearly three acres of garden space. Its fifteen thousand rose bushes and arches highlight over eight hundred rose varieties. Each June, Rose Weekend is celebrated, with guests treated to tours and family-friendly activities.

Viola papilionacea, Viola sororia

Violet

Symbols & Meanings

The violet can symbolize many things; at its core it is a flower of duality. Modesty, faithfulness, innocence, and love are all associated with the violet; however, in many cultures, it is associated with grief and mourning.

It is said that when you dream of violets, they represent a life filled with joy and love. Because of their symbolic connection to everlasting love, violets are often gifted to newlyweds.

Violets are one of the earliest spring bloomers, hence they are the birth flower for February.

The violet's sweet aroma is often associated with relaxation, as it has a calming effect and helps alleviate stress.

Botany

The violet—also known as the wood violet, common blue violet, meadow violet, and woolly blue—is a short-stemmed perennial with white, blue, or purple flowers and glossy heart-shaped leaves. Native to eastern North America, it grows in well-drained, shady habitats.

The violet offers an unimposing yet powerful floral presence. Often considered a weed, the violet can be found in lawns and gardens as well as in woods, thickets, and stream banks.

History

The violet has been well-known and referenced since the time of ancient Greece, most notably by the poet Sappho (610 BCE–570 BCE), who often mentioned violets in her work. The violet has been used for centuries as an herbal remedy for different ailments.

Home

The violet is the state flower of Illinois, Rhode Island, New Jersey, and Wisconsin.

Uses

The violet's edible flowers are used to make wine and to sweeten dishes; the leaves are sometimes used in salads and cooked greens. The flowers adorn dishes and cheese boards, adding a bright pop of purple, and its leaves offer a boost of vitamins A and C. Beauty, taste, and health benefits all packed into one gorgeous little blossom!

Pollinators

Violets are visited by a wide range of pollinators including bumble bees and butterflies. Fritillary caterpillars, such as the great spangled fritillary and variegated fritillary, depend on these plants for their survival. Violets also serve as food for wild turkeys, rabbits, deer, livestock, mourning doves, bobwhites, and white-footed mice.

Folklore

Violets have a deep and longstanding folkloric and literary history. Their connection to death, grief, and mourning permeates many cultures.

The violet reigns supreme in mythological underworlds. In ancient Rome, the violet represented both grief and purity of heart and therefore was often used to adorn the graves of innocent youths departed too soon. In Lithuanian mythology, the violet is associated with their god of the underworld, Poklius, as well as Old Prussia's god of the underworld, Patulas.

Persephone, queen of the underworld in Greek mythology, also had a penchant for the violet. When she was taken by Hades, she was in the middle of harvesting flowers, the violet being among the bunch. As she reigned as queen of the underworld, the violet was the earthly pleasure she longed for the most.

In literature the violet is mentioned in several of Shakespeare's works. It is a symbol of both faithfulness and death in *A Midsummer Night's Dream*, *Twelfth Night*, *Hamlet*, and *Pericles*.

Recipe

VIOLET REFRIGERATOR JELLY

Foraging wild violets in the warm sun can be done right in your own backyard! Once you spot and properly identify these precious purple flowers, simply remove them from the stem, rinse, and allow them to dry on a paper towel before using in this scrumptious recipe. To create the perfect bite of earthy, floral richness, pair violet jelly with homemade butter on warm biscuits or sourdough.

MAKES 16 OZ [475 ML]

2 loosely packed cups (about 1½ oz [40 g]) purple wild violet flowers

One 32 oz [945 ml] lidded jar, sterilized

2¼ cups [540 ml] boiling water

½ powder packet [25 g] powdered pectin

1½ cups [300 g] granulated sugar

2 Tbsp freshly squeezed lemon juice

One 16 oz [475 ml] lidded jar, sterilized

1. To make the violet tea concentrate: Rinse the violets by gently swirling them in a bowl of water or rinsing them in a salad spinner to remove any soil. Transfer the violets to the 32 oz [945 ml] jar and pour in the boiling water. Cover with the lid and let the violets steep for 30 minutes.

2. Meanwhile, in a small bowl, mix the powdered pectin with 2 tablespoons of the sugar. Set aside.

3. Strain the tea through a sieve into a saucepan, pressing to extract as much liquid as possible. Discard the blossoms.

4. Stir the pectin-sugar mixture and the lemon juice into the tea. This will turn the tea from blue to purple.

5 Bring the violet tea to a boil, stirring constantly.

6 Stir the remaining sugar into the violet tea. Boil for 1 more minute, stirring constantly.

7 Remove from the heat and skim off any foam with a spoon.

8 Ladle the jelly into the 16 oz [475 ml] jar, close securely with the lid, and pop into the fridge. The jelly will keep, refrigerated, for 3 weeks.

Leucanthemum vulgare

Oxeye Daisy

Symbols & Meanings

The oxeye daisy is often associated with innocence, purity, and new beginnings. This summertime bloom heralds the return of warmer seasons, a welcome harbinger of abundant sunshine and the joy of nature. Its abundance of white petals around a yellow center is a familiar icon of childhood, evoking deep nostalgia for many.

Botany

The oxeye daisy is a widespread flowering plant native to Europe and Asia but now found worldwide. It reaches a height of around 3 ft [91 cm] and has flowers like those of a typical daisy. It grows in a variety of plant communities and is famous for its highly successful propagation—in fact, the oxeye daisy is considered an invasive species in *forty* countries. It spreads by seeds—a single mature plant can produce over twenty thousand seeds in a season—and by its creeping rhizomes, or rootstocks, spreading underground to form a vast subterranean system.

Etymology

The word "daisy" derives from the Old English *dæges eage* ("day's eye"), referring to the way the flower opens its petals in the morning to capture sunlight and closes them at night.

The French word for daisy, *marguerite*, derives from the Greek word for pearl, *margarite*.

History

The oxeye daisy was first recorded in the early Iron Age in Britain. It was introduced to the United States in the 1800s and propagated quickly. The plant has a medicinal history, too: Traditionally, it was used to treat various health problems, such as coughs and asthma. The flower heads are still used to make a delicious herbal tea.

Pollinators

The oxeye daisy is a top contributor to the health and wellness of local ecosystems through its support of a diverse variety of pollinators, including beetles, ants, butterflies, bees, bumble bees, mason bees, sweat bees, wasps, and flies, who all flock to the flower's nectar and pollen.

Pop Culture

The oxeye daisy is the famous flower used in the classic "loves me, loves me not" game, where petals are picked to divine the lover's fate.

Folklore

The oxeye daisy is deeply rooted in femininity. In worldwide mythology it is designated the flower of choice for the Norse goddess Freya and the Greek goddess of the moon, Artemis.

The oxeye's womanly power also appears with lesser divinities, like the nymph Belides, who transformed herself into a daisy to escape the unwanted advances of Vertumnus, the Roman god of seasons and gardens.

502
The FARMER'S DAUGHTER Flowers
502
502
E OHIO ST
BUS STOP
WELCOME
SEED
SEED

FLOWER SHOP

The Farmer's Daughter Flowers

Pittsburgh, Pennsylvania

All flower shops have a certain magic about them, but this one holds a special place in our hearts because, well, Lauren, one of the authors of this book, is the owner.

Picture this: the jingle of the front doorbell, phones ringing, music, the smell of coffee, and laughter. When you step into The Farmer's Daughter Flowers, the curated, handpicked gifts gracing the storefront and the fresh fragrance of flowers awaken your senses and infuse you with a sense of joy. Antique chandeliers hanging from the ceiling, a Parisian-style stem bar, and an array of houseplants fill this city space, making it feel like a hidden gem in the middle of metropolitan hustle and bustle. While the shop is located in the city of Pittsburgh, we have our own boutique flower farm where we grow our own stems to feature in our floral designs. Locally sourced at every opportunity, from our farm and other local farms to our customers' vases.

A FLORIST'S MISSION

Flowers are *only* given to bring joy. They are there to ease pain, brighten up a day, celebrate a life lived, celebrate love, and bring peace into the home. How lucky it is to work with a medium that is meant to bring joy alone. Not to mention, being able to experience them from seed to bloom provides endless inspiration. Working with flowers is a beautiful life.

—Lauren Work, owner, The Farmer's Daughter Flowers

Part V

The Oddities

Though we tried to gather our most beloved flowers, this collection represents only a peek into the vastness of the flower world. The Oddities section is an ode to the rare and wondrous flowers that make us wonder how we are lucky enough to share a lifetime with these inspiring plants. Each flower in this visual showcase is paired with its name and the symbolic meaning it holds.

Rare and Unusual Flowers

CHECKERED LILY
(Fritillaria meleagris)
Devotion and humbleness

BLACK BAT FLOWER
(Tacca chantrieri)
Elegance and strength

PASSIONFLOWER
(Passiflora)
Peace and harmony

ROTHSCHILD'S SLIPPER ORCHID
(*Paphiopedilum rothschildianum*)

Rarity and uniqueness

THE CORPSE FLOWER
(*Amorphophallus titanum*)

Intrigue and drama

HIMALAYAN BLUE POPPY
(*Papaver betonicifolium*)

Purity and happiness

Poisonous Flowers

CLEMATIS (*Clematis*)

Wisdom and aspiration

POISON HEMLOCK
(*Conium maculatum*)

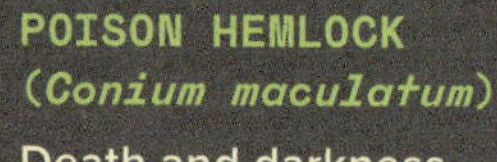

Death and darkness

BLEEDING HEART

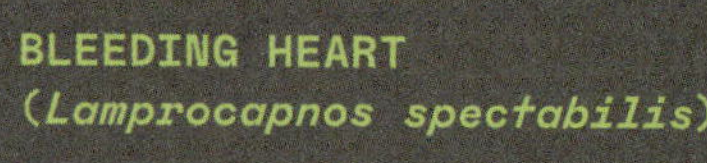

(*Lamprocapnos spectabilis*)

Love and heartbreak

DEADLY NIGHTSHADE
(*Atropa belladonna*)

Danger and beauty

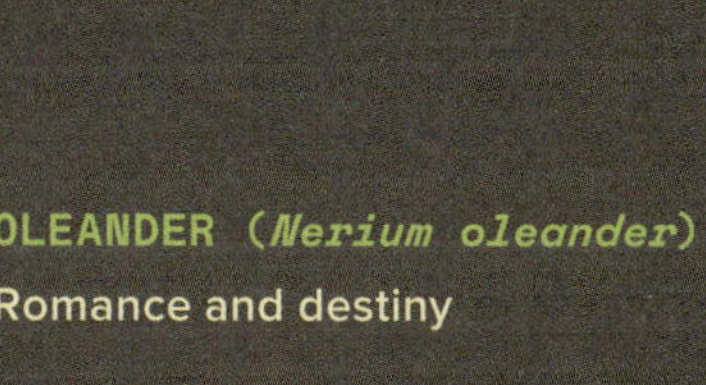

OLEANDER (*Nerium oleander*)

Romance and destiny

LILY OF THE VALLEY
(*Convallaria majalis*)

Innocence and motherhood

Night Bloomers

JASMINE (*Jasminum officinale*)

Purity and awakening

ANGEL'S TRUMPET (*Brugmansia suaveolens*)

Health and vivacity

GARDENIA (*Gardenia*)

Affection and refinement

EVENING PRIMROSE
(Oenothera biennis)
Fondness and love

MOCK ORANGE
(Philadelphus coronarius)
Sweetness and joy

MOONFLOWER *(Ipomoea alba)*
Mystery and intuition

Desert Flowers

DESERT GLOBEMALLOW
(*Sphaeralcea ambigua*)
Reflection and ambition
QUEEN OF THE NIGHT CACTUS
(*Epiphyllum oxypetalum*)
Transience and ephemera
FIREWHEEL
(*Gaillardia pulchella*)
Endurance and adaptability

Water Flowers

JAPANESE IRIS
(*Iris ensata*)

Bravery and wisdom

MARSH MARIGOLD
(*Caltha palustris*)

Renewal and joy

SACRED LOTUS (PINK)
(*Nelumbo nucifera*)

Rebirth and awakening

WATER POPPY
(*Hydrocleys nymphoides*)
Tranquility and transformation

WATER HYACINTH
(*Pontederia crassipes*)
Sincerity and grace

CARDINAL FLOWER
(*Lobelia cardinalis*)
Distinction and vibrancy

CLOSING

In the stillness of nature, there's a quiet wisdom that gently guides us, one we often overlook in our busy lives. Flowers, in all their stages, mirror the natural rhythm of the seasons and remind us of the importance of honoring each one.

We begin by planting seeds, whether in the soil or in our lives, with care and intention. We nurture them through patience, effort, and perseverance. When they finally bloom, we're reminded that every bit of work was meaningful, and it's essential to pause and appreciate the beauty of our effort. But just as the earth knows when to rest, so must we. After the planting and the blossoming, there is a season of stillness. This is the time to pause, reflect, and replenish.

Mother Nature offers a timeless lesson: to value our efforts, to celebrate the beauty that follows, and to give ourselves the grace to rest. It's in this cycle that we find true balance and renewal.

Recommended Reading

ROSE

Appleyard Flowers. "Flowers Named after Royals." https://www.appleyardflowers.com/flowerdiaries/flowers-named-after-royals/.

Blooming Haus. "Flowers Named after the Royal Family." https://bloominghaus.com/shop/flowers-named-after-the-royal-family/.

Country Living. "Royal Flowers: Queen Elizabeth's Favorites." https://www.countryliving.com/uk/homes-interiors/gardens/g41191600/royal-flowers-queen/.

Encyclopaedia Britannica. "Rose." https://www.britannica.com/plant/rose-plant.

Gardenia. "*Rosa* (Rose): Plant Information & Care." https://www.gardenia.net/genus/rosa-rose.

Kew Gardens. "Roses at Kew: A History and Their Uses." https://www.kew.org/plants/roses.

Ludwig's Roses. "The Rose in Myths & Legends." https://www.ludwigsroses.co.za/latest-news/the-rose-in-myths-legends/.

University of Illinois Extension. "History of Roses." https://extension.illinois.edu/roses/history-roses-0

SUNFLOWER

American Sunflower Association. "Sunflower History." https://www.sunflowernsa.com/all-about/history/.

Kansas Farm Food Connection. "Why Kansas Is the Sunflower State." https://kansasfarmfoodconnection.org/spotlights/why-kansas-is-the-sunflower-state.

Save Our Monarchs. "Spotlight on Eastern Black Swallowtail Butterflies." https://www.saveourmonarchs.org/blog/spotlight-on-eastern-black-swallowtail-butterflies.

Smithsonian American Art Museum. "Clytie: Sunflower Mythology." https://americanart.si.edu/artwork/clytie-20026.

The Tarot Guide. "The Sun Card Meaning." https://www.thetarotguide.com/the-sun/.

University of Illinois Extension. "Sunflowers Symbolize Summer, Elevate Garden Displays." 2022. https://extension.illinois.edu/blogs/flowers-fruits-and-frass/2022-08-12-sunflowers-symbolize-summer-elevate-garden-displays-and.

University of Missouri Extension. "Sunflowers: History, Production, and Benefits." https://extension.missouri.edu/publications/g4290.

Victory Garden Gal. "Sunflowers Give Power to the Victory Garden." https://victorygardengal.com/sunflowers-give-power-to-the-victory-garden/.

DAHLIA

American Dahlia Society. "Dahlia Myths: Part 1." 2018. https://dahlia.org/wp-content/uploads/2018/01/Dahlia-Myths-Part-1.pdf.

American Dahlia Society. "Dahlia Myths: Part 3". 2018. https://dahlia.org/wp-content/uploads/2018/01/Dahlia-Myths-Part-3.pdf.

American Dahlia Society. "Dahlia Nomenclature and Brief History." https://www.dahlia.org/docsinfo/articles/dahlia-nomenclature-and-brief-history/.

Dahlias.com. "Annual Dahlia Festival." https://www.dahlias.com/our-farm/annual-dahlia-festival/.

Flower Meaning. "Dahlia Flower Meaning." https://www.flowermeaning.com/dahlia-flower-meaning/.

International Society for Horticultural Science. "Dahlia as an Important Plant for Human Health." https://www.ishs.org/ishs-article/1288_15#:~:text=Dahlia%20is%20an%20important%20plant,and%20lows%20cholesterol%20and%20triglycerides.

Kew Gardens. "Dahlia." https://www.kew.org/plants/dahlia.

Landuum. "Dahlia: The Representative Flower of Mexico." https://www.landuum.com/en/history-and-culture/dahlia-the-representative-flower-of-mexico/.

IRIS

American Herbal Products Association (AHPA). "Herbs in History: Iris." https://www.ahpa.org/herbs_in_history_iris.

Botanical.com. "Irises: History and Uses." https://www.botanical.com/botanical/mgmh/i/irises08.html.

Forest Service, USDA. "Iris: Wildflowers of North America." https://www.fs.usda.gov/wildflowers/beauty/iris/index.shtml.

Gardens Illustrated. "Skellorn Irises & Iris Breeding." https://www.gardensillustrated.com/features/skellorn-irises-iris-breeding.

Missouri Integrated Pest Management (IPM). "Iris: A Brief History." 2014. https://ipm.missouri.edu/meg/2014/6/Iris-A-Brief-History/.

Travel France Online. "Fleur de Lys: French Monarchy's Emblem." https://www.travelfranceonline.com/fleur-de-lys-french-monarchys-emblem/.

World History Encyclopedia. "Iris: Goddess of Rainbows and Mythology." https://www.worldhistory.org/Iris/.

POPPY

Bee Life. "Do Bees Like Poppies? https://www.beelife.org/do-bees-like-poppies/.

British Legion. "The Poppy and Remembrance." https://www.britishlegion.org.uk/get-involved/remembrance/about-remembrance/the-poppy.

California Department of Fish and Wildlife. "California Poppy Conservation & History." https://wildlife.ca.gov/Conservation/Plants/California-Poppy.

DEA Museum. "Opium Poppy: Nature's Addictive Plant." https://museum.dea.gov/exhibits/online-exhibits/cannabis-coca-and-poppy-natures-addictive-plants/opium-poppy.

Farmer's Almanac. "Poppy Flower Facts, Symbolism, and Gardening Tips." https://www.farmersalmanac.com/poppy-flower-facts-symbolism-and-gardening-tips.

Rails, Eric. "History of the Poppy Flower." *Earth.com.* https://www.earth.com/earthpedia-articles/history-of-the-poppy-flower/.

MAGNOLIA

Arbor Hill Trees. "Magnolia Tree Facts." https://arborhilltrees.com/blog/magnolia-tree-facts/.

Caerhays Estate. "History of Magnolias." https://visit.caerhays.co.uk/the-estate/the-gardens/history-of-magnolias/.

Carter, Maria. "The Heartbreaking Real-Life Story Behind *Steel Magnolias*." *Country Living.* https://www.countryliving.com/life/entertainment/a44128/steel-magnolias-30th-anniversary/.

Mississippi Encyclopedia. "Southern Magnolia." https://mississippiencyclopedia.org/entries/southern-magnolia/.

Mississippi State Library. "Mississippi State Symbols: Tree and Flower." https://msstate-exhibits.libraryhost.com/exhibits/show/mssymbols/treeandflower.

WISTERIA

A to Z Flowers. "Wisteria." https://www.atozflowers.com/flower/wisterias/.

Bonsai-En. "Wisteria Species Guide." https://bonsai-en.shop/blogs/tree-species-guide/wisteria-species-guide.

Flower Meaning. "Wisteria Flower Meaning." https://www.flowermeaning.com/wisteria-flower/.

Missouri Botanical Garden. "*Wisteria sinensis.*" https://www.missouribotanicalgarden.org/PlantFinder/PlantFinderDetails.aspx?taxonid=280632

Mommy Poppins. "Sierra Madre Wistaria Festival 2024." 2024. https://mommypoppins.com/los-angeles-kids/event/events/sierra-madre-wistaria-festival-2024.

My Flora Guide. "Ode to the Wisteria: The Bardini Garden in Florence." 2021. https://www.myfloraguide.com/en/2021/04/16/ode-to-the-wisteria-the-bardini-garden-in-florence/.

Oxford University Herbaria. "*Wisteria sinensis* Profile." https://herbaria.plants.ox.ac.uk/bol/plants400/Profiles/WX/Wisteria.

CAMELLIA

American Camellia Society. "*Camellia sinensis*: Backyard Tea." https://www.americancamellias.com/education-and-camellia-care/the-camellia-family/camellia-sinensis-backyard-tea.

Flower Meaning. "Camellia Flower Meaning." https://www.flowermeaning.com/camellia-flower-meaning/.

This Is Alabama. "Beauty and History: The Story Behind Alabama's State Flower, the Camellia." January 2, 2019. https://www.thisisalabama.org/beauty-and-history-the-story-behind-alabamas-state-flower-the-camellia/.

Tuscan Trends. "Visiting the Camellia Festival of Sant'Andrea di Compito and Pieve di Compito." March 13, 2023. https://www.tuscantrends.com/visiting-the-camellia-festival-of-santandrea-di-compito-and-pieve-di-compito/.

DOGWOOD

Dogwood Garden Club. "Legends of the Dogwood." https://dogwoodgardenclub.org/legends-of-the-dogwood/.

Druids Garden. "Sacred Trees in the Americas: Flowering Dogwood (*Cornus florida*)—Magic, Medicine, and Mythology." May 2, 2021. https://thedruidsgarden.com/2021/05/02/sacred-trees-in-the-americas-flowering-dogwood-cornus-florida-magic-medicine-and-mythology/.

Greenwood Nursery. "Best Tips for Gardening with Bluebirds." https://www.greenwoodnursery.com/blog/best-tips-gardening-bluebirds.

Vinton Messenger. "What Is the Legend of the Dogwood Tree?" https://vintonmessenger.com/vinton-history-museum-what-is-the-legend-of-the-dogwood-tree/.

CHAMOMILE

Almanac. "Flower Meanings: The Language of Flowers." *The Old Farmer's Almanac.* https://www.almanac.com/flower-meanings-language-flowers.

American Herbal Products Association (AHPA). "Herbs in History: Chamomile." https://www.ahpa.org/herbs_in_history_chamomile.

Missouri Botanical Garden. "*Matricaria chamomilla* (German Chamomile)." https://www.missouribotanicalgarden.org/plantfinder/PlantFinderDetails.aspx?taxonid=277347.

National Center for Biotechnology Information (NCBI). "Chamomile: A Herbal Medicine of the Past with a Bright Future" (2010). https://pmc.ncbi.nlm.nih.gov/articles/PMC2995283/.

Penn State Extension. "Culinary Herbs Are Good for Beneficial Insects, Including Pollinators." May 21, 2024. https://extension.psu.edu/culinary-herbs-are-good-for-beneficial-insects-including-pollinators.

ScienceDirect. "Chamomile: Pharmacology, Toxicology, and Pharmaceutical Science." https://www.sciencedirect.com/topics/pharmacology-toxicology-and-pharmaceutical-science/chamomile.

Tilth Alliance. "Growing Herbs for Pollinators." *Garden Almanac.* https://tilthalliance.org/resources/growing-herbs-for-pollinators/.

World History Encyclopedia. "Ra (Egyptian God)." https://www.worldhistory.org/Ra_(Egyptian_God)/.

LAVENDER

Almanac. "Lavender: How to Plant, Grow, and Care for Lavender." *The Old Farmer's Almanac.* https://www.almanac.com/plant/lavender.

Flower Meaning. "Lavender—Meaning, Symbolism, and Colors." https://flowermeanings.org/lavender/.

Guest, Linda. "MU Extension Research on Lavender Finds Options for Missouri Growers." *University of Missouri Extension.* https://extension.missouri.edu/news/mu-extension-research-on-lavender-finds-options-for-missouri-growers.

Kew. Plants as LGBTQ+ Symbols. Royal Botanic Gardens, Kew. https://www.kew.org/read-and-watch/plants-LGBTQ-symbols.

"Lavender." *Encyclopedia Britannica*. https://www.britannica.com/plant/lavender.

Missouri Botanical Garden. "*Lavandula angustifolia* (English Lavender)." https://www.missouribotanicalgarden.org/PlantFinder/PlantFinderDetails.aspx?taxonid=281393

New Crops & Organics. "Lavender: History, Taxonomy, and Production." *North Carolina State University Extension.* https://newcropsorganics.ces.ncsu.edu/herb/lavender-history-taxonomy-and-production/.

Provence-Alpes-Côte d'Azur. "Lavender Festival." https://provence-alpes-cotedazur.com/en/things-to-do/events/lavender-festival/.

Simply Beyond Herbs. "10 Lavender Meanings and Symbolism." https://simplybeyondherbs.com/lavender-meanings/.

The Practical Herbalist. "Lavender: Love-Inducing Protector." https://thepracticalherbalist.com/advanced-herbalism/lavender-love-inducing-protector/.

UIC Heritage Garden. "Lavender (*Lavandula*)." *University of Illinois, Chicago, Heritage Garden.* http://heritagegarden.uic.edu/lavender-lavandula.

University of California Agriculture & Natural Resources. "History and Uses of Lavender." https://ucanr.edu/sites/EDC_Master_Gardeners/files/168086.pdf.

U.S. Lavender Growers Association. "Lavender Varieties." https://www.uslavender.org/index.php?option=com_content&view=article&id=73:lavender-varieties&catid=24:lavender-101&Itemid=138.

You're History. Folklore in My Garden: Lavender. July 19, 2014. https://yourehistory.wordpress.com/2014/07/19/folklore-in-my-garden-lavender/.

DANDELION

Flower Meaning. "The Dandelion: Its Meanings & Symbolism." https://www.flowermeaning.com/dandelion-flower-meaning/.

Gardening Know How. "Dandelion Plant History & Facts." https://www.gardeningknowhow.com/tbt/dandelion-plant-history-facts.

Maine Organic Farmers and Gardeners Association. "Ten Things You Might Not Know about Dandelions." https://www.mofga.org/resources/weeds/ten-things-you-might-not-know-about-dandelions/.

Reconnect with Nature. "5 Things You Didn't Know About Dandelions." https://www.reconnectwithnature.org/news-events/the-buzz/5-things-dandelions/.

The Practical Herbalist. "Dandelion History, Folklore, Myth, and Magic." https://thepracticalherbalist.com/advanced-herbalism/dandelion-history-folklore-myth-and-magic/.

HIBISCUS

Almanac. "Hibiscus: How to Plant, Grow, and Care for Hibiscus Flowers." *The Old Farmer's Almanac*. https://www.almanac.com/plant/hibiscus.

Flower Meaning. "Hibiscus Flower Meaning." https://www.flowermeaning.com/hibiscus-flower-meaning/.

Hindu of Universe. "Why Do Hindus Worship Flowers?" https://hinduism.hinduofuniverse.com/why-do-hindus-worship-flowers/.

Maui Nui Botanical Gardens. "Maʻo Hau Hele *(Hibiscus brackenridgei)*." https://mnbg.org/hawaiian-native-plant-collection/mao-hau-hele-hibiscus-brackenridgei/.

Narayan Seva Sansthan. "Unveiling the Mystique of Goddess Kali: Significance, Legends, and Symbolism." https://www.narayanseva.org/unveiling-the-mystique-of-goddess-kali-significance-legends-and-symbolism/.

National Library Board Singapore. "Hibiscus (*Hibiscus rosa-sinensis*)." https://www.nlb.gov.sg/main/article-detail?cmsuuid=58f5f860-ace3-4819-82c5-be7d2477855f.

National Park Service. "Home of Pele Field Trip." https://www.nps.gov/common/uploads/teachers/lessonplans/Home%20of%20Pele%20field%20trip.pdf

Pacific Horticulture. "The Hibiscus Revolution." https://pacifichorticulture.org/articles/the-hibiscus-revolution/.

State Symbols USA. "Pua Aloalo: Hawaiian State Flower." https://statesymbolsusa.org/symbol-official-item/hawaii/state-flower/pua-aloalo.

University of Hawaii. "*Hibiscus brackenridgei*." *College of Tropical Agriculture and Human Resources, Hawaiian Native Plant Propagation Database*. https://www.ctahr.hawaii.edu/hawnprop/plants/hib-brac.htm.

University of Illinois College of Agricultural, Consumer & Environmental Services. "Huge Hibiscus Flowers Are a Garden Standout." *Illinois River Hort*. July 28, 2017. https://extension.illinois.edu/blogs/ilriverhort/2017-07-28-huge-hibiscus-flowers-are-garden-standout.

MARIGOLD

Almanac. "Marigolds: How to Plant, Grow, and Care for Marigolds." *The Old Farmer's Almanac*. https://www.almanac.com/plant/marigolds.

Almanac. "October Birth Flowers." *The Old Farmer's Almanac*. https://www.almanac.com/october-birth-flowers.

Desert Botanical Garden. "How Marigolds Became a Symbol for Día de Muertos." https://dbg.org/how-marigolds-became-a-symbol-for-dia-de-muertos/.

Extension Minnesota. "Marigolds." University of Minnesota. https://extension.umn.edu/flowers/marigolds.

Flower Meaning. "The Marigold: Its Meanings & Symbolism." https://www.flowermeaning.com/marigold-flower-meaning/.

Our Lady of the Lake University. "Marigolds and Día de los Muertos Traditions." *OLLU Library Guides.* https://libguides.ollusa.edu/diadelosmuertos/marigolds.

Piedmont Master Gardeners. "Marigold: More Than a Pretty Face." https://piedmontmastergardeners.org/article/marigold-more-than-a-pretty-face/.

COMMON MILKWEED

Johnny Butterflyseed. "The Monarch Festival in Bethel, Maine: A Celebration of Nature's Majestic Migrants." https://www.johnnybutterflyseed.com/2023/08/15/monarch-festival-bethel-maine/.

Marble Crow. "Milkweed Folklore and Magic." 2003. https://marblecrowblog.com/2023/09/06/milkweed-folklore-magical-properties/.

Penn State Extension. "Monarchs and Milkweed." https://extension.psu.edu/monarchs-and-milkweed.

US Department of Agriculture. "Butterfly Milkweed. *Plant Guide*. https://plants.usda.gov/DocumentLibrary/plantguide/pdf/pg_astu.pdf.

PURPLE CONEFLOWER

Marble Crow. "Echinacea Folklore and Magic." 2003. https://marblecrowblog.com/2023/08/23/echinacea-folklore-and-magical-properties/.

Missouri Botanical Garden. "*Echinacea purpurea.*" https://www.missouribotanicalgarden.org/PlantFinder/PlantFinderDetails.aspx?kempercode=c580#:~:text=Echinacea%20purpurea%2C%20commonly%20called%20purple,%2Deyed%20Susans%20(rudbeckias).

MOUNTAIN LAUREL

Brooklyn Botanical Garden. "Mountain Laurel: A Shade Tolerant Native." *Garden Stories.* 2016. https://www.bbg.org/article/mountain_laurel.

Plant Native. "Native Mountain Laurel." 2025. https://theplantnative.com/plant/mountain-laurel/.

Wellsboro Area Chamber of Commerce. "Pennsylvania State Laurel Festival." https://www.wellsboropa.com/index.php/pa-state-laurel-festival.

WILD COLUMBINE

Almanac. "How to Grow Columbines." 2025. https://www.almanac.com/plant/columbine.

Arnow, Olivia. "The Legend and Lore of the Columbine." *PostIndependent.* 2024. https://www.postindependent.com/news/local/outdoors-the-legend-and-lore-of-the-columbine/; https://santaclaracommunity.org/scco/13th-annual-wildflower-celebration-may-4th-awbrey-park/.

Greg App. "Symbolism and Benefits of the Wild Columbine." 2024. https://greg.app/wild-red-columbine-benefits/.

University of Texas Lady Bird Johnson Wildflower Center. "*Aquilegia canadensis.*" https://www.wildflower.org/plants/result.php?id_plant=AQCA.

University of Texas Lady Bird Johnson Wildflower Center. "Eastern Red Columbine." https://www.wildflower.org/plants/result.php?id_plant=aqca.

VIOLET

Cole, Nellie. "Folkdays: Violets." May 7, 2021. https://nelliecole.com/2021/05/07/folkdays-violets/.

NC State Extension Gardener Plant Toolbox. "*Viola sororia.*" https://plants.ces.ncsu.edu/plants/viola-sororia/.

Save the Bee. "Five of the Most Bee Friendly Flowers to Plant in Your Garden." https://savethebee.org/5-of-the-most-bee-friendly-flowers-to-plant-in-your-garden/

Simply Beyond Herbs. "Violet Flower Meaning and Symbolism: Ultimate Guide." April 14, 2025. https://simplybeyondherbs.com/violet-flower-meaning/.

Wigington, Patti. "Violet Magic and Folklore." 2020. https://www.pattiwigington.com/violet-magic-and-folklore/.